THE QUEST FOR REALITY

The Quest for Reality

Bohr and Wittgenstein
Two Complementary Views

STIG STENHOLM

OXFORD
UNIVERSITY PRESS

OXFORD
UNIVERSITY PRESS

Great Clarendon Street, Oxford, OX2 6DP,
United Kingdom

Oxford University Press is a department of the University of Oxford.
It furthers the University's objective of excellence in research, scholarship, and education by publishing worldwide. Oxford is a registered trade mark of Oxford University Press in the UK and in certain other countries

First published 2011
First published in paperback 2015

Impression: 1

Published in the United States of America by Oxford University Press
198 Madison Avenue, New York, NY 10016, United States of America

British Library Cataloguing in Publication Data
Data available

Library of Congress Cataloging in Publication Data
Data available

ISBN 978–0–19–960358–9 (Hbk.)
ISBN 978–0–19–872910–5 (Pbk.)

Printed and bound in Great Britain by
Clays Ltd, St Ives plc

CONTENTS

1 Prelude: The modern stance 1
- 1.1 A time of change 1
  - 1.1.1 Outline of argument 3
- 1.2 Certainty lost 6
  - 1.2.1 Algorithmic or analogical thinking? 8
  - 1.2.2 Pictures in the head 10
  - 1.2.3 Language and mind 10
  - 1.2.4 The whiff of truth 12
- 1.3 Declaration of intent 14
  - 1.3.1 Role of philosophy 14
  - 1.3.2 Stranger in a strange land 16
  - 1.3.3 The metaphysical observer 18

2 Twilight of the gods 22
- 2.1 The dawn of darkness 22
- 2.2 Cultural backgrounds 24
  - 2.2.1 Wittgenstein's Vienna 24
  - 2.2.2 Bohr's Copenhagen 31
  - 2.2.3 Kierkegaard as symbol of the times 33

3 The view from Copenhagen 48
- 3.1 Bohr the master 48
  - 3.1.1 Life and position 48
  - 3.1.2 Background and influences 50
  - 3.1.3 References relating to Bohr 52
  - 3.1.4 The life and opinions of Høffding 52
- 3.2 Bohr's legacy 57
  - 3.2.1 Bohr's conceptual analysis 57
  - 3.2.2 Bohr's view of physics 59
  - 3.2.3 The late views, 1958–1962 66
- 3.3 Rosenfeld's approach 68
  - 3.3.1 The life of Léon Rosenfeld 68
  - 3.3.2 Rosenfeld on epistemology 69
- 3.4 Copenhagen summarized 75
- 3.5 Bohr's pragmatism 76
  - 3.5.1 William James and pragmatism 77
- 3.6 Russian sensitivity 81

4 Epistemological interlude 86
4.1 Background considerations 86
4.1.1 Philosophy and science 86
4.1.2 Words and worlds 87
4.2 The human position 88
4.2.1 Animal awareness 88
4.2.2 Consciousness itself 90
4.2.3 The sensational ambience 91
4.2.4 In the back of my mind 92
4.3 Ways of representation 94
4.3.1 Constituting reality 94
4.3.2 The conversing human 96
4.3.3 Capacity for symbolic forms 98
4.3.4 The use of formal structures 100
4.4 Role of formalism 101
4.4.1 Symbols used and misused 101
4.4.2 Tests and classifications 104
4.4.3 Overcoming paradoxes 107
4.5 Formal procedures in physics 109
4.5.1 The test (forming a judgement) 109
4.5.2 What entities can we posit? 112
4.5.3 The complement: what we do not assert 113
4.5.4 Is there a total world? 114
4.5.5 Classical approach 115
4.5.6 Quantum approach 116
4.6 Physical applications 119
4.6.1 Numbers of objects 119
4.6.2 Space 121
4.6.3 Time 124
4.7 What the theoretical structures tell us 127
4.7.1 Quantum theory revisited 127
4.7.2 Truth and consequences 128

5 Wittgenstein enters the scene 130
5.1 Life and work 130
5.1.1 Background for Wittgenstein 131
5.2 Influences on Wittgenstein 136
5.2.1 Reading in philosophy 136
5.2.2 The role of Weininger 137
5.2.3 Role of Boltzmann 142
5.2.4 Interview with von Wright 146
5.3 Stages of Wittgenstein's philosophy 148
5.3.1 The early Wittgenstein 148
5.3.2 Wittgenstein's influence 153
5.3.3 The later Wittgenstein 155

5.4 What can we really know? 160
5.4.1 Moore on knowledge 160
5.4.2 Wittgenstein on certainty 160
5.4.3 von Wright on certainty 162

6 Shaky foundations 164
6.1 In logic we trust? 164
6.1.1 Quest for certainty 164
6.1.2 The frail temple of mathematics 165
6.1.3 Interpretation of mathematics 166
6.1.4 The beauty of formalism 167
6.2 Mathematics as a sovereign world 171
6.2.1 Logic and language 172
6.2.2 Language 172
6.2.3 On objects 174
6.2.4 Signs and symbols 176
6.2.5 Procedures of mathematics 177
6.2.6 Reasons and causes 179
6.2.7 Wittgenstein on foundations of mathematics 179

7 Physics interface 183
7.1 Tractatus and the classical world 183
7.1.1 The binary description of reality 186
7.2 Quantum physics turns to language 187
7.3 Wittgenstein and physics 189
7.3.1 Science and method, method of science 189
7.3.2 Physical objects 191
7.3.3 Language and symbolism 192
7.3.4 Phenomena 193
7.3.5 Experiment 195

8 Philosophical consequences 196
8.1 The right to freedom 196
8.2 Pragmatism at stake 197
8.2.1 Summary of pragmatism 197
8.2.2 Pragmatism in the modern world 198
8.3 Husserl's last stand 199
8.3.1 Life and fate 199
8.3.2 Husserl's main ideas 200
8.3.3 The crises in European science 201

9 Metaphysics and reality 204
9.1 What can we know? 204
9.1.1 Embedding knowledge into reality 204
9.1.2 On what there is and is not 205

9.2 Beyond the veil of endeavors 206
9.2.1 Leeway for beliefs 206
9.2.2 Role of religion 208

10 Concluding epilogue 213

References 216

Index 221

1

PRELUDE: THE MODERN STANCE

1.1. A time of change

Philosophie dürften man eigentlich nur dichten[1]

The centre cannot hold[2]

Medieval man lived in a universe of certainty. The Church delivered truth, nay it was the only truth. If one queried science or issues of conduct, the scholars could provide an answer. This was always to be found in Aristotle or the Church fathers respectively. Deviations from this received judgement were incinerated in the auto-da-fé, consuming both the individuals as well as their writings. No deviating answers emerged because the questions could not be asked.

The first break in this bulwark was caused by science. The movement initiated by Galileo Galilei could not be stemmed and eventually deprived the Church of its authority in scientific matters. The process was slow but its outcome was inevitable. Sustained by progress in technology, modern physics and chemistry proved their worth.

Concomitantly with the scientific developments, the reformations ended the supreme sovereignty of the Catholic Church and opened up the freedom of the individual to think. The French Revolution finally put its seal on modernity; even in the matter of administration, the popular opinions of the crowd could no longer be neglected. The modern society is still living as the heir to these movements.

But physics did not provide the ordered world expected of it; nor did any brand of democracy lead to the *greatest happiness for the largest number of people.* All that happened was that we were thrown into a state of perplexity, which only grew with increasing knowledge. The increase of knowledge certainly did not increase happiness.

In science, the mastering of heat engines and electricity forged the foundations for most of modern technology. It was, consequently, expected that the phenomena of nature could be catalogued and chained. Things did not turn out this way; the ultimate theory of matter and radiation, quantum theory, took the form of an abstract set of recipes, which gave no basis for a visualizable picture of everything. Still the theory has been profoundly successful when applied to real

[1] L. Wittgenstein, *Vermischte Bemerkungen.*
[2] W. B. Yates, *The Second Coming.*

phenomena; both technology and fundamental understanding have benefitted enormously. But it has not solved any riddle of our lives nor revealed the essence of the empirical world.

When philosophy aimed at certainty it turned to mathematics. Surely, it was taught, the result of a calculation is necessarily true. The statement "The circumference of a circle is π times its diameter" has to be an absolute truth. But no, the arbitrariness of Euclidean geometry made the statement undecided at best. With a closer investigation, more difficulties emerged: How do we assign a numerical value to the diagonal line and the curved line of the circle itself; how do we compare these numbers? What is that mysterious symbol "π" and what is its status of being; how are its digits determined and what is their distribution?

In addition to the questions of defining the mathematical concepts and their mutual relations, there turned out to be unavoidable features of incompleteness. Not all theorems can be proved and not all numbers can be computed. These results do, of course, refer only to formal processes in closely defined systems, but they suggest a need to doubt the truth and relevance of mathematical manipulations.

The consequence of these complications turned out to be the insight that all knowledge must be captured in words belonging to our common language. Because this is not a closed formal structure, it is dependent on human life in the everyday world. It contains limitations deriving from our limited intellect and historical developments as social beings. There can be no universal language game describing all of reality including human consciousness. The use of language works only in limited situations and in incomplete ways.

Ludwig Wittgenstein started his thinking by trying to provide a complete description of all that can be stated uniquely in language. The rest cannot be discussed and thus lies outside of communication. He thought that he had finished the task of philosophy, which thus had reached its ultimate end. Later he recognized that language is more intricate than he had assumed. His subsequent work was to be based on communication within limited systems, considered as language games in which concepts have their meaning. Even the activity of mathematicians falls into this category of games.

In physics, Niels Bohr came to be the advocate of a new approach to the physical investigation of reality. He recognized that the theories of physics also derive their communicative power from the use of language, even if we need to extend it with the tools of mathematics. This is then to be taken as complementing our common language. The theory is purely formal, and communication has to be conducted in terms of classical concepts, because these apply to the world where we live. The only way to make the exchange of physical information unambiguous is to extract classical pictures from the processes of nature. According to Bohr, the full description from initial preparation to final recording of the outcome, is the basic *physical phenomenon.* This is the unit of description, just as Wittgenstein's language game is his playground for meaningful communication.

I have singled out Wittgenstein and Bohr for a comparison, because they represent the best examples of the modern approach to the fuzziness resulting from progress in understanding the human position in the world. Each one had to replace apparent order and certainty by an understanding based on limited concepts in constant flux.

Bohr and Wittgenstein worked in very different fields and locations. I certainly do not claim that they influenced each other directly; they may not even have known about each other. Their parallel endeavors derive from the cultural atmosphere of all civilized Europe. They reacted in their separate characteristic ways to the prevailing zeitgeist. Hence their achievements are symptoms not mutual causes. And still, their influence on their contemporaries became unforeseeably large. This is part of the magic they radiated into the realms of their activities.

Bohr and Wittgenstein are complementary in treating human communication from abstract formalism to empirical methods. But both realized that the problem lies not in nature nor in our scientific methods but in our use of language. This implies that they even tend to use identical ways to express themselves.

Niels Bohr:

> We are suspended in language in such a way that we cannot say what is up and what is down.

Ludwig Wittgenstein:

> We are struggling with language.

This defines the tools, the goal can only be defined by a poet [1]:

> What are the roots that clutch,
> what branches grow out of this stony rubbish?
> Son of man.
> You cannot say, or guess, for you know only
> A heap of broken images, where the sun beats.

1.1.1. Outline of argument

The aim of my writing is to consider the breakdown of a certain world view during the twentieth century. This is mirrored in the treatment both of science and philosophy. As these form the foundation of the human position in the world, a major reorganization of the body of knowledge had to take place. I have chosen Bohr and Wittgenstein as the main actors to represent this revision. My conclusion is that the modern synthesis is far from satisfactory. By leaving the story unfinished in its present state, I hope to provoke a renewal of the discussion and the eventual emergence of a reformed clarity and understanding. Reality is to be left as it is, but the human mind is also to be left as it is. The problem is to find a way to effect a harmonious combination of these.

The content of this book is essentially one argument. However, because it encompasses such a broad range of human activities, it will consist in various pieces which together, hopefully, suggest the unified picture I aim at.

Chapter 1: The modern stance

This outlines the motivation I have had for undertaking this work. I also argue that, acknowledging the remarkable progress of science, it is essential to dissect the modern world view from the aspects of physics. The physicist may be the outsider here, but he may also bring new light on a confusing situation. In order to combine the thinking in physics with that of philosophy, I choose to deal with the lives and achievements of Bohr and Wittgenstein in some detail.

The total analysis is based on my personal experience and subjective feelings; it thus remains a fiction: *The Myth of Physics and the World.* Only when definite conclusions are someday reached, will it be the time for formal analysis and tracing of roots. For the moment we are not yet there.

Chapter 2: Twilight of the gods

Here I describe the loss of intellectual certainty characterizing the nineteenth century. Vienna as well as Copenhagen was dominated by the cloud of despair characterizing the period between the Napoleonic Wars and the disastrous First World War. In spite of all progress, the former catastrophe created an atmosphere designed to anticipate the second one. In retrospect, many symptoms of this were to be seen, but the most influential one was, no doubt, the emergence of the Dane Søren Kierkegaard. In his native Copenhagen he initiated the existentialist movement influencing all Europe, and he was well known in Vienna at Wittgenstein's time.

Chapter 3: The view from Copenhagen

This summarizes the life and work of Niels Bohr. It presents his view of modern physics and possible influences on him, and the impact of his activities. In particular, I discuss the role of his philosophy mentor Høffding and colleague Rosenfeld. Finally I compare Bohr's approach to the *pragmatism* of William James.

Chapter 4: Epistemological interlude

In this Chapter, I summarize my personal view of existence as a human being in empirical reality. Much is speculative and many things are highly subjective. However, I consider it intellectually honest to inform the readers about my own standing. This bias may induce me to superimpose my views on those I present later. It should be up to the reader to decide to what extent Bohr and Wittgenstein support my approach. This chapter treats: epistemology, consciousness, language, and the methods of science. All these form central issues in the later discussions, and this introductory story should be conceived as a preliminary background. In conclusion it states that what we experience of nature, is never enough to lay down the basic elements of reality and their essences. Metaphysics is empirically underdetermined but essential for a harmonious world view.

Chapter 5: Wittgenstein enters the scene

Here I summarize the life and achievements of Wittgenstein. I trace the philosophical background of his activity, stressing especially the role of Boltzmann. Wittgenstein never became his student but was, obviously, strongly influenced by him. I also contend that the opinions of Boltzmann were very similar to those of pragmatism, even if it is highly unlikely that Boltzmann knew any works by William James. Here I also summarize the conventional description of Wittgenstein's thoughts, the early and the later ones. There are, in fact, lots of similarities between them: Wittgenstein's problems remained largely the same, but the early certainty was replaced by an extensive skepticism and feeling of helplessness in view of the immense problems offered by life and logic. The chapter is concluded by the discussion on *Certainty* originating from lectures delivered by the philosopher George Edward Moore.

Chapter 6: Shaky foundations

Here I summarize the effort to reach absolute certainty in mathematics. This endeavor ended in failure, because no absolute truths can be found even in formal systems. The underdetermination of all such systems is taken to imply that not even the best theory can reach an absolute truth. In view of this, Wittgenstein devotes a lot of discussion to the very activity of doing mathematics: What can we know and what can we do even without a solid foundation? If the world of mathematics is posited, what life can live and develop in this world?

Chapter 7: Physics interface

It is argued that the *Tractatus* is an analogy of the classical description of physical reality. We know that this had to be replaced by the much more ephemeral quantum theory. It is argued that many of the features of Wittgenstein's later work could have been applied to quantum physics, if only he had known about this. The argument at some points refers directly to modern physics in a way Wittgenstein could not have done; he did not know modern science.

Chapter 8: Philosophical consequences

We return to the issue of pragmatism and its relevance to modern physics. It is argued that Husserl's *Phenomenology* is the last attempt of nineteenth-century thinking to impose conventional order on the realm of philosophy and its manifestations in our empirical picture of reality. I argue that this enterprise failed, and after this the European continental philosophers confined themselves to the world of human life and its written products. Thus *existentialism* has given up the attempt to understand and control the real environment. This separates them unequivocally from the philosophy of science in any of its possible forms.

Chapter 9: Metaphysics and reality

Here I discuss the implications deriving from the epistemology that has served science so well. Our technical mastery of the empirical world still does not uniquely determine either the ontology of the ultimate reality or the potential interdependences of it elements. What we can know we may know, but our understanding allows, nay demands, that there are things we cannot know. Thus we are free to populate these posited worlds as we like, assuming only that we do not create inner contradictions into our fantasies or external conflicts with empirical facts as we know them. This situation leaves open the opportunities even to incorporate religious arguments. To avoid misunderstanding: religious concepts are in no way implicated by our knowledge but, on the other hand, it cannot exclude them. It is pointed out that Wittgenstein was apt to refer to God, especially in his early writings.

Chapter 10: Concluding epilogue

This chapter comments on the opinions presented. I concede that many arguments presented have been expressed earlier; only the combination of issues may offer a novel perspective. However, a detailed declaration of relations to other works would only have proved lengthy and boring. The writing itself is supposed to convey the necessary information to the reader.

I am aiming at exhibiting an impression of the worlds of reality as seen from a position of modern science. The result is multifariously different from a strictly analytic formulation. I am fully aware that the text satisfies neither the formal requirements of academic philosophy nor the deductive logic of physical theories. It is a hazy view, but it suggests to me the existence of a thing of beauty, the human participation in reality.

Consequently there is no attempt to include a complete list of references. If each statement and each quotation had been referred to its proper place in the literature, the text would have been cluttered with notes and remarks, seriously hampering the reading. Thus only essential sources are indicated.

1.2. Certainty lost

The quest for certainty in philosophy has left a feeling of dissatisfaction. Here I am trying to present a vision or a program. Thus I have to speak a lot about things one cannot speak about according to the philosophers. But I have to, this is part of the problem.

The progress of philosophy during this century has been most impressive. However, when all is said and done, and a lot has been said indeed, we are left rather unsatisfied. The grandiose program to systematize mathematical knowledge ended up in an incompleteness theorem, and the ambitious positivistic effort to construct the world from sense data collapsed under its own immensity [2]. Philosophy itself seems to have abandoned all high expectations and in various postmodernistic forms joined the arts in a struggle for ingenuity and originality but forgetting goals related to reality and relevance. This will not do; if philosophy is not to become a fringe of knowledge dealing in verbal acrobatics only, it must return to its original tasks of making sense of our best understanding of reality and elucidating our role as humans in the world of being.

This century is characterized by a huge expansion in our scientific knowledge. Chemistry and physics have entirely altered our everyday surroundings as well as our view of the Universe. One may question which aspect is more significant for human life, but that does not change the situation. Molecular biology tells us how we become what we are, and the theory of evolution why. Empirical science is also gaining more and more information about the working of our brains, and the questions about the character and mode of existence of consciousness are again allowed in a scientific discourse [3]. Many details are still obscure, but there is no denying that the world where the educated man lives today is vastly different from that of our forefathers only a hundred years ago.

But this huge leap in our knowledge has not really reformed the process of philosophical thinking. The modern analytic school started as an attempt to understand mathematics as being the most rational of human activities. The result seems to be that its complexity and very essence are too incomprehensible to allow any universal conclusions. The attempt failed, but the methods and goals of this approach have survived as "the method" of philosophy. It has been applied to other fields of so-called exact sciences: astronomy, physics, and chemistry. These produce numbers, thus creating an illusion of exactness. However, the competent practitioner knows well how many approximations go into even the best of theoretical predictions.

Most of the traditional philosophy of science is based on the classical disciplines of physics, which are easier to comprehend than the modern quantum physics and field theory. These are very abstract, and their very mode of existence is dubious. We simply do not know enough! When such theories have been analyzed by philosophers, very sensational conclusions have been reached: they can be made to support both anti-realist interpretations of knowledge [4] and mystic views of the unity of everything [5]. Little useful insight has emerged, and it seems that these theoretical structures cannot be analyzed fruitfully by the methods of present philosophy.

However, philosophy must remain a human enterprise. It must follow the development of our best knowledge and its impact on society and the individual. Thus it must take a phenomenological starting point and admit the facts of existing reality. If we want to create a philosophical understanding of our world, we must start with science, even if philosophy may have other goals too, which need a different point of view. The methods developed to understand mathematics have been found not to work here; they did not really work for mathematics either. The certainty envisaged by Descartes was not to be found. The fact that physics deals in numbers does not make it a part of mathematics. It is an ongoing attempt to describe reality using the tools of our language, which are hopelessly inadequate to the task even if, as Bohr puts it, they are supplemented by the language of classical physics [6]. There is no safety in numbers, nor in anything else. Philosophy cannot make progress if it does not abandon its requirements of rigor and certainty and joins the natural sciences in their uncertain and groping search for partial truths and reasonable conclusions. The progress in science has proved that this can be done without giving up the ideals of increasing verisimilitude and growing comprehension.

As active scientists we tacitly assume the reality and existence of the objects of our inquiry. No philosophy can change that.

But how can we know that reality exists? At the dawn of the twentieth century, Ernst Mach [7] explicitely denied the reality of the then emerging microscopic world:

> The moment we begin to operate with mere things of thought like atoms and molecules, which from their very nature can never be made the objects of sensuous contemplation, we are under no obligations whatever to think of them as standing in spatial relationships which are peculiar to the Euclidean three-dimensional space of our sensuous experience.

However, science presupposes some brand of reality. As the archmaterialist Lenin succinctly expresses it:

> The scientists will also have to answer this question unhesitatingly; and they do invariably answer it in the affirmative, just as they unhesitatingly recognize that nature existed prior to man and prior to organic matter.

The Russian attitude to Quantum Theory will be discussed below.

Another answer is that we all live as if a real world existed. Even the most ardent anti-realist philosopher steps aside when a fast car is approaching; he does not stop to analyze his sense impression or the mode of being of the car. How do we know that other humans exist? Because we see them behave as we do and describe their own experience just as we do ours. These statements cannot be made certain; they are based on common sense. But any other sense is far less common.

We may thus start our discourse from the observation that human individuals live in the world, act in purposeful ways, and react meaningfully to external occurrences. We may infer that they are engaged in an activity we call thinking, even when we do not know all that this implies.

1.2.1. Algorithmic or analogical thinking?

The dominating influence computers have on our society today has led to a tendency to pattern all understanding on the manner of operation of the digital data processor. This holds not only in the sciences but equally in linguistic and social inquiries. All intellectual activity is supposed to be modelled on the algorithmic way a computer deals with data. But the computer is basically a very primitive contraption, and it may be a mistake to take its construction as the paradigm of our theoretical effort to understand reality.

The mode of operation of a computer has been formalized in the concept of a *Turing device.* This is the universal computer, which can emulate the operation of all classical ones. As such it can be subjected to logical analysis, and the limitations of logical systems apply to its capabilities. In particular, the incompleteness theorems associated with the name of Gödel have their computational counterparts: not even all well-formulated tasks are computable. This occurs even for some problems where we, the humans, seem to know the answers. Penrose [8] takes this as proof that the human mind does not operate algebraically; there are intuitive truths which can not be computed on a Turing device.

It seems that Penrose's conclusion is going too far; what is known and what is not in an algebraic system depends on the formal structures encompassed. All proofs work only within these systems, and their authority to say something about reality is questionable. Only a Platonic interpretation of necessary relations between formal systems and external reality could confirm such authority on them, and the validity of such an interpretation is just part of the problem we are trying to elucidate. The Penrose argument is clearly circular.

But the fact remains that humans are good at making propitious decisions in a complex and ever-changing world. The human mind is an excellent machine for drawing useful conclusions from insufficient information. It seems highly implausible to derive this property from an algorithmic model of the working of the mind.

There is, however, a different way of obtaining solutions to problems, the analog computation. This has been nearly forgotten in this age of digital processing, but its characteristic feature is to get acceptable overall solutions without being able to achieve any feats of extreme accuracy. A good analog model is fast and retains the essential features of the situation, but it cannot iterate through many steps without losing its precision. It requires continuous feedback from the environment and updating of its inputs. Analog methods are, however, known to be able to solve problems not amenable to numerical treatments.

Here it may be objected that an analog device can be emulated on a digital computer. This is true but essentially irrelevant. The digital solution is based on a mathematical formulation of the problem, which can then be translated into a numerical algorithm. It does not prove that the original method was algorithmic. For instance, a complicated boundary value problem can be solved simply by making a wire frame of the shape of the boundary and dipping it into a soap solution. I fail to see how this procedure could be considered algorithmic.

The seemingly eerie ability of the mind to surmise correct solutions may derive from an analog activity. Then we can retain a materialistic view of thinking without hitting the limitations of algorithmic processes. The brain consists of a complicated, highly interconnected, and extremely parallelized data processing unit. In this case the distinction between digital and analog processing blurs. Its activity may well be based on manipulating the information in an analog manner by pattern matching, associative combinations, and emotional steering. This may allow fast and efficient decisions which would work for all practical purposes. It would provide what we are used to call intuition.

1.2.2. Pictures in the head

The approach outlined above brings back the metaphysical nightmare of ideas as images in our minds. It has to assume that we possess structures in the nervous system which are maps of various features of reality. The mapping may be rather complicated; our conscious categories of space and shape may well be coded totally differently in the mind. The images are not only descriptions of the objects and their relations and properties, but they may also contain much more abstract webs of relations connecting them to our social and professional situation. All features of reality which have a bearing on our well-being and activities must be mirrored in the structure of the mind.

We have no theoretical model of such imaging systems, but to me they seem no less likely than any corresponding digital maps. And some representation of reality in the mind is necessary to explain our behavior in the world. Brain physiological research gives ever more evidence of the complex interconnection between our bodily and mental activities. On the other hand, Jungian psychology has provided the concept of Archtypes in the mind [9]. These are common structures we humans possess, they are genetically conditioned, developed by language, and enforced by society. They may well be manifestations of the innate structures of the mind, as these emerge through the methodological tools of analytic psychology.

How is it possible that our minds contain images of these complex features of reality? Such a correspondence between our internal and external worlds may seem purely metaphysical until we remember that we are what we are because of biological evolution. Humans, including their mental apparatus, are the endpoints of a Darwinian process, which has served to make the individuals fit to live in their environment. The details of this process are too complex to be fully understood, but there prevails a general consensus about the basic facts.

It seems reasonable to think that the evolution process has impressed visions of the surrounding reality on the human mind. The ability to learn must contain the same trends, so that education and growing up add further images of the complex systems we encounter in nature and society. Both the history of mankind and the trial and error of the individual life serve to eliminate grossly incorrect conceptions of reality. The learning process is highly flexible allowing totally new skills to emerge. There can be no genetic disposition to learn driving a car in a congested urban area. Furthermore, if the human mind can learn to feel at home in N-dimensional topological spaces and solve tasks there, how can we doubt the existence of images of surroundings much more necessary for survival?

1.2.3. Language and mind

In some sense we have returned to Kant, the description we have of reality is determined by the categories of our mind. But we have a different emphasis, the categories utilized by our knowledge are shaped by reality as experienced by humankind. They are adapted by evolution and learned by living. We have an incomplete, fractured, but highly interconnected representation of external reality in our minds.

When we act, we can draw directly on our inner images, but if we wish to communicate about them we have to resort to language. This is an attempt to map the inner images on the structures provided by language. Thus a verbal description is twice removed from the reality it tries to convey; no wonder it is imperfect. It is understood only on the basis of an underlying common life experience and a principle of charity; in similar situations people are expected to have similar impressions, obtain similar associations, and have similar reactions. Here any cultural differences may introduce difficulties; one need not think of the anthropologist in an isolated tribe, but even the discussions between physicists and biologists are sometimes bizarre enough.

This brings us to Wittgenstein's idea of language as a picture of the world [10]. He abandoned this view himself, which action may well be justified by the difficulty of catching the multifarious features of reality in a linguistic framework. The situation changes, however, if we assume that the verbal expression tries to be a picture of an inner representation of reality. This is already a symbolic representation, even if we do not know how the coding operates. Thus it makes much more sense to interpret the language pictures of Wittgenstein's *Tractatus* as representations of such inner images of the speaker. These constitute his impression of the external world, but they are also conditioned by all other features making up the individual speaking. Any information conveyed depends on a common language, a common life experience, and a common biological inheritance. It still holds that: "Wovon man nicht sprechen kann, darüber muss man schweigen."[3] Inside a human there is much one cannot speak about.

Our approach thus admits that we build up our image of the environment using the categories available to the mind, and we can only communicate them to others if our language contains the means to do so. Where both Kant and Wittgenstein go wrong is in their conviction that these limitations remain constant. Evolution and training can teach us to experience things not accessible to our forefathers, and languages can develop to contain expressions not earlier needed. The progress of science has led to a desire to incorporate large parts of mathematics into the common language, but this only serves to illustrate the process. Even the logically offensive quantum theory can be the subject of meaningful and rational discourse.

The images we have in our minds have been shaped by reality and thus they must contain genuine features of the structures and interrelations in the external world. We have thus direct access to such information, and utilizing all the innate understanding life and evolution have given us, we may be in a position to start the endeavor envisaged by Plato and Descartes, namely to try to understand the nature of reality, the workings of our minds, and the fact that they depend on each other. This is what philosophers call the *problem of intention*; how can

[3] What we cannot speak about, we must pass over in silence.

features of the mind represent objects in the outer world? Much relevant data have been collected by science, but a new "method" to approach these problems has not even started to emerge.

Surprisingly, the view we have reached seems to agree well with both ancient philosophy and modern counterintuitive features of quantum theory. Even in those simple situations where quantum physics can be applied theoretically and tested empirically, the phenomena contain amazing nonlocal correlations and holistic dependences, which are in direct conflict with the simpleminded classical view of reality. Philosophers have long warned us that this view is an illusion, but quantum theory cannot be taken as the final verification of the unreality of the external world. After all, the concepts and rules of the theory have been extracted from purely empirical circumstances; they have been forced on us by the world itself. To take the theory as proof of its unreality constitutes a circular argument.

The conclusions proposed in this text sound highly mystical. If that is what they are, so be it; mysticism is just a label and need not be taken in a pejorative sense.

But the arguments used in this book are clearly metaphysical. If such arguments are to be excluded, then I have nothing to say. It is, however, my strong conviction that no deeper understanding of reality can be gained without entering metaphysics. In fact, I would even say that those arguing that they can manage without metaphysical assumptions are simply deceiving themselves. The problem is that their assumptions are either hidden or simplistic, mostly both. We cannot talk about reality, empirical or theoretical, without metaphysical presumptions. Nature is neither an abstract mathematical structure nor a human construct. I strongly feel that, if a future philosophy is to contribute to our understanding of reality and man's place in it, one cannot continue to neglect those aspects of our common human endeavor which are expressed here. In this book I am trying to capture the essence of the human experience as it has emerged during the twentieth century. Both physics and philosophy have met with immense progress and bitter disappointments. But from contortion agility is born.

1.2.4. The whiff of truth

Truth is one of the most used (and misused) concepts of philosophy. The situation is simple in a well-ordered world, where each object is identifiable and all its properties are uniquely verifiable. We may then list all objects of the world and the relations (*Sachverhalten* according to Wittgenstein) they take part in. The truth of a statement is decided by a simple correspondence test; it is true if it is in agreement with the state pertaining to the real world.

However, such a simple situation as outlined above does not seem to be attainable even in the realm of classical epistemology. Modern linguistics and especially modern physics seem to make the simple situation described an illusion. There may be no fundamental objects constituting the world and their defining characteristics may be inseparable from their role as actors in forming the totality of being. Thus *truth* needs to be reconsidered; there is no absolute concept of truth that can be read off the appearance of reality even including the human interpreter.

The logical concept of "truth" was institutionalized by Alfred Tarski, who formulated the rules needed to introduce truth into formal languages pertaining to logic. This work has become a cornerstone for the analysis of truth, but it depends on an ambience of a formal language, and thus it carries no implications for the analysis of the real existence of the world. It helps us check the consistency of the concept of truth when one is offered to us, but it gives no method to look for it. Donald Davidson has attempted to extend this to the ordinary language used in everyday communication, but there seems to be no consensus that his approach exhausts the topic.

The present analysis derives from the starting point that truth may mean different things in different types of discourse. The opposite of "true love" may be "false love" but this concept has but little in common with "false money" or "faithless husbands". *Truth comes in flavours.* The obvious question is then: how many flavours are there? Instinctively one may say that there are infinitely many, but that statement is not very helpful. In this writing I will try to analyze what I believe is a main division. Most instances of the application of the attribute "true" falls under one of these categories.

Like flavours, truth comes in three categories: sweet, salty, and bitter. I analyze them one by one:

Sweet truth

Mathematical truth which can be proved within a system. Form a sentence according to the syntax of your system and logically independent. As a consequence of Gödel's theorem, any such sentence can be assigned a provable truth value, true or false; in some cases only if the axiomatic system or the tools for proofs are suitably extended. The impact of such truth on reality is questionable. Outside its formal framework it lacks consequences. There seems to be no limit to the amount of mathematical truth possible. It is not known if, in a Platonic sense, there exists a maximal set of mathematical truth. Personally I doubt it.

Salty truth

Situations which "everybody" agrees to be manifest. These are the direct observational sentences that at one time were supposed to constitute the "sense impressions" from which a world is to be built. We agree that there is an apple on the table, I see that the bottle is empty, and my technical device shows the numeral 12.5. The truth here is that we see the apple, we notice the emptiness of the bottle, and the meter shows 12.5; these states of the matter prevail. The experience is true, because we (somehow) manage to eliminate the possibility of an illusion. I also leave out the question whether all observers agree on the same experience, but for the present purpose this may be assumed. What these "truths" say about the external world is not true or false in a formal sense. They require an interpretation which falls outside the present category. Take the statement "The bottle is empty." I say that at dinner, the other guests may or may not agree, but as an absolute statement about the world, it is not unambiguously relevant. Is it really a bottle, and if it is, is it empty, and how can we test if it is so? All these questions depend on material circumstances, and it is not obvious that the concepts utilized have meaning in the appropriate context.

Bitter truth

Statements which acquire their warranted assertibility from the prevailing level of scientific knowledge. In Newton's time it was true to state that the motion of planets is determined by his law of gravitation; today this is approximately true only. At Bohr's time, the atomic nucleus consisted of protons and neutrons, but then no true statements about quarks were possible. That quantum objects can appear as waves or particles is true, but the question about their nature cannot be formulated at the present level of science. Each meaningful scientific statement about the world may be true or false within its limits of applicability, but it is not, and presumably will never be, absolutely true or false. Water is a liquid, but it is also H_2O, depending on the context of the use of the statement. Even universal concepts like *Laws of Nature* cannot be true or false; they can only be valid as approximately verified.

1.3. Declaration of intent

1.3.1. Role of philosophy

An essential question for all humans is the meaning of our existence and the existence of the external world of experience. Traditionally such ponderings have belonged to the hazily defined discipline of philosophy. However, during the progress of scientific inquiry, philosophy has had to adjust to the emerging mass of accumulated facts, because it can obtain its legitimacy only by resting on accepted truth. On the other hand, it remains controversial to decide what constitutes established truth. Thus we are back at a question belonging to the realm of philosophy. Thus empiricism and human interpretation of its products have progressed hand in hand, causing conflicts but also creating concepts to be utilized by the other side.

At the dawn of the twentieth century, the emergence of modern physics was strongly influenced by philosophical considerations. Both Einstein and Heisenberg produced their revolutionary results under the influence of positivistic trends. *Only that can exist which allows for an operational interpretation.* Later developments have passed beyond this puristic approach, but the associated physical achievements have remained valid. So something true must have emerged from the argument.

Later developments have estranged philosophy and the developments of modern physics. Both parties are here, in my opinion, to blame. Physics has been blinded by its own success; the progress in experimental technique and the massive achievements of commercial applications have made the physicists overlook the absurd world view offered by quantum theory.

Philosophy has been of no help here, it has shunned the hard questions concerning reality and turned to logistic formalism, philosophy of everyday phenomena or pure linguistic pettiness. This way academic philosophy has turned away from its duty to build a conceptual framework for the exploration of physical reality. The situation has become such that leading physicists have turned their back on all philosophy. A characteristic example is the detailed argumentation of the Nobel Laureate Steven Weinberg [11]:

> I do not even mean to deny all value to the philosophy of science, which at its best seems to me a pleasing gloss on the history and discoveries of science. But we should not expect it to provide today's scientists with any useful guidance how to go about their work or about what they are likely to find.

The weird character of quantum knowledge has, however, initiated a movement to understand the whys and hows of the grossly successful quantum theory. Today it is again permissible to ask questions about its foundations and the character of physical knowledge. Philosophy is back on the stage; whether it can help us to push physics beyond its present stalemate remains to be seen.

We need to reconsider both the epistemological stand of scientific inquiry and the ontological implications of the ever-so-successful modern physics. Fundamental analysis needs to be resurrected while considering the progress in empirical science. Whatever reality is, its empirical manifestations must be mirrored in our thinking. This is no preconceived view of what we expect to find, but a part of the method of our discourse.

If we want to progress beyond our present state of human understanding, we must evade all indoctrination from earlier thinking. We must consider the play anew. We must regard the empirical world, its impression on our minds, and the relation between these as a natural phenomenon to be described and studied. In this book I attempt to lay the foundations for such an activity.

It may be claimed that such audacity indicates hubris in a high degree. Especially as philosophers are as lax to admit any influence from natural science as ever the physicists with respect to philosophy. Our multifariously eminent philosopher G. H. von Wright repeatedly stresses [12]:

> I find particularly distasteful the kind of "holiday thinking" in which scientists—some of great stature in their professional field—sometimes indulge in order to express their nonprofessional views on the "big questions": on the nature of mind and matter, on the meaning of life, or on the divine hand of nature. At the risk of being accused of professional conceit, I should say that whereas there are good examples of successful amateurs in science, I cannot think of any in philosophy.

This opinion may be countered by the question "Who is a professional philosopher?" Many of the great masters have not set out by a formal training in their discipline but have entered the brotherhood only on the merits of their work. *Post festum* they are thus judged not to be amateurs. Von Wright's damnatory remark is presumably directed towards Eddington, whose contribution to philosophy did not stand high with von Wright. This I learned from a personal discussion with him many years ago.

On the other hand, the famous physicist Richard Feynman once wrote [13]:

> To prepare oneself to be the theoretical physicist who will some day find the key to these strong interactions, it might be thought that a full knowledge of all these complicated formulations would be necessary. This may be so, but the exact opposite may also be so; it may be necessary to stay away from the corners where everyone else has already worked unsuccessfully.

He published this in 1961, before the emergence of the "standard model" which is now so successful. This succeeds just because it oversteps the boundaries of what was taken to be in the theory in 1961. Maybe we need such a re-evaluation of the whole concept of a philosophy of science.

The need to philosophize by a doer and not by a talker was acknowledged by Ludwig Boltzmann when he in 1903 was inaugurated to the chair in Natural Philosophy in Vienna. He then had but scant formal merits for such an appointment, his main production being pure physics up to then. He, however, had to succeed Mach, who had emerged as a most influential character in the interpretation of classical science. In his inaugural lecture, Boltzmann justifyied his new career as follows [14]

> Indeed if Moltke was made a member of the historical division of the Berlin Academy because he had not written but made history, perhaps I too was chosen not because I had written about logic, but because I belong to a science that offers the best opportunities for daily practice in strict logic.

No meek claim to put forward, but historical developments have proved Boltzmann to be right, even if he personally did not reap the full benefits of his thinking. I will return to his role in the scientific history of the twentieth century.

Maybe, in this new century, it is time for the natural scientists to survey the philosophical landscape with radically novel spectacles. This implies that much of received wisdom has to be rejected. I am not writing a textbook in academic philosophy; nor am I drawing strict truths from accepted premises. I am neither analyzing nor axiomatizing. I am committing the ultimate sin, I am drawing conclusions about reality from my own subjective experience of the world.

1.3.2. Stranger in a strange land

Let us try to capture the landscape we encounter in modern science. We have to establish the conceptual tools we choose to accept in order to have some starting point, *Ex nihil nihilo fit.* The rest is to find the appropriate questions for the enquiry. "A whole course of philosophy can consist entirely of questions" according to a remark by Wittgenstein.

"We live in the real world." Thus far we can all agree. However, all words in this manifesto may be the object of doubts: Who are the "we" making a statement about ourselves? In this context "to live" is more than to exist, which is a dubious concept in itself. Our "living" cannot be inferred from mere "thinking" as we cannot show that our experience of thinking is not itself part of that "thinking". Because we cannot conclude anything about the experience of a nonthinking entity, the Cartesian argument carries no conviction.

The role of "living" implies feats beyond mere "existence". To live is to act on experience, draw conclusions from environmental influences, and react back on the environment. A "living" entity stands in a causal relation to its environment. But this introduces a new dubious term "causal". When is an act or an occurrence causally implied by some state of affairs? Causality stands in need of a definition in terms of its usage, which does not simplify the enigma.

Finally, what is the environment? The "real world" grasped as the environment is the third dubious term in the manifesto? The world is everything there is. But this again brings in the question of existence but now on a cosmic scale.

The summary above is an itinerary for millennia of philosophy. Its progress has served to eliminate the most absurd interpretations of the concepts above, but no final verdict has emerged from the process.

Our task is to continue the discussion, but we have to start by establishing which rules of conduct we accept for the proceedings.

The totality of reality is like the dark interior of a strange continent. We enter it and encounter phenomena weird to our experience from our well-ordered homeland. The forest is full of strange fruits and surprising movements. We cannot even easily judge what are features of the landscape and what are manifestations of foreign life. As explorers we should not apply our preconceived concepts to our experience. We should only describe what we see, look for regularities and species of living entities. Thus we may gradually build up a physical and biological understanding of our novel environment.

Some occurrences may be recognized as intelligent life. As devoted anthropologists, we try to communicate with these, however strange their language and habits are. Only when we have established such communication, can we try to understand the world view and rites of these beings.

Paul Feyerabend recognizes the fact that to a cosmic observer our science and its processes may be absurd [15]. The visiting anthropologist has to establish communication within our culture before he can make sense of our behavior and the accompanying declarations:

> Returning from his field study to his own conceptions and his own language, such as English, an anthropologist often realizes that a direct translation has become impossible and that his views and the views of the culture to which he belongs, are incommensurable

with the "primitive" ideas he has just begun to understand. Naturally he wants to give an account of these ideas in English, but he will be able to do so only if he is prepared to use familiar terms in strange and novel ways. He may even have to build an entirely new language-game out of English words, and he will be able to start his explanation only when his language-game has become fairly complex. Now we know that almost every language contains within itself the means of restructuring large parts of its conceptual apparatus.

Feyerabend claims that the received view of scientific progress is invalid. New results do not emerge in a logical way from the collected empirical evidence. They are born in a dialectic activity derived from intuition, coercion and communication. But this is a process requiring time and effort with built-in possibilities for misunderstanding and human deviousness. He describes the process as follows:

> One must use sentences for which no clear rules for usage are yet available. Just as a child starts using words without yet understanding them, who adds more and more uncomprehended linguistic fragments to his playful activity, discovers the sense-giving principle only after he has been active in this way for a long time—the activity being a necessary presupposition of the final blossoming forth of sense—in the very same way the inventor of a new world view (and the philosopher of science who tries to understand his procedure) must be able to talk nonsense until the amount of nonsense created by him and his friends is big enough to give sense to all its parts.

Feyerabend is usually taken to advocate a nihilistic view of scientific progress: "Anything goes." His argument may, however, be given a different interpretation: "So you claim that progress in science is achieved in a rational way. Then I can present a rational argument for the view that science progresses irrationally." *Reductio ad absurdum*?

We should, however, take seriously Feyerabend's description of the creation of new scientific frameworks. When such a process is carried to conclusion, we have a valid scientific description of the phenomena studied, but before that point we may encounter seemingly irrational and absurd situations. Is that where quantum mechanics stands today?

In order to attempt progress in this situation, we must establish what we regard as acceptable foundations. This implies establishing some description of ourselves as experiencing subjects.

1.3.3. The metaphysical observer

As stated above, one view of philosophy is to regard it as a tool to achieve true knowledge about the world and the role of mind in this. However, in order to derive anything from anything, we must have a solid foundation from which to progress. We have to take something for granted before we start the argument. Cartesius took the "thinking" as certain and later logical world constructions have attempted a start from pure *sensational impressions* as objects of direct acquaintance.

Carnap and his followers hoped to logically construct the world as we know it. The process was never completed, and today it seems to be abandoned as just as hopeless as the behavioristic approach to mind. The exactly opposite view is inherited from Plato: The objects constructed from pure thought must necessarily manifest themselves in the world albeit in a distorted shape. This approach is widely recognized as invalid except by modern mathematical physicists. For them, abstract groups and other formal structures are naively taken to offer a valid foundation for description of experiments. However, the only proved case is the general theory of gravitation. Here Einstein's pure thought created a framework apparently describing empirical reality adequately. The emergence of quantum physics was much more contorted and less amenable to analysis.

There are two contrary approaches to the individual as an experiencing entity. First the world of experience is created by the mind and all concepts relating to the external world lack reality. Such an idealistic interpretation of knowledge cannot be refuted, but it carries little conviction and leads to absurdities when pushed to its ultimate limits. Secondly, a purely materialistic view is that the brain is a complicated informational processor, and the subjective feeling of consciousness is an epiphenomenon without independent existence. Even this approach lacks conviction; how are we supposed to convince an epiphenomenon that this is all it is?

In spite of all progress in brain science, consciousness is just as far from an explanation as it ever was. Even if we clinically find that a certain signal in the brain indicates "red" and another indicates "horse", how are we to explain the experience of a red horse? Measurements can only correlate physical quantities but no systematic approach has been able to convert them into subjective experiences. We cannot satisfactorily explain consciousness from any theory we know. It also seems that no progress in the physiology of the brain is able to explain it. Here we encounter a situation analogous to the one in physics: We can successfully apply quantum theory to all observed processes of physics. But even so, we cannot explain the emergence of the everyday "classical" world where we live. At best, we may show that it is possible without inconsistency, but its evident necessity has remained beyond reach.

Let us see what implications we may find for the starting of philosophical enquiry. We are all conscious of being conscious. Thus it seems impossible to "explain consciousness" to itself. If that is so, let us take the consciousness as a primitive concept to start from: the entity performing our analysis is a mind receiving impressions and processing these, possibly deciding to act on the information received or the conclusions drawn. This simple change of view avoids many problems; the character of mental experience and the existence of free will pose no enigmas. I do not claim that this move is unavoidable but only that it delivers fruitful results. The ultimate foundations for building reality can only be known to gods and not to humans.

So what kind of information can the conscious subject take cognizance of? On the one hand, we have the sense impressions entering through the sense organs.

In the world as experienced these are based on repeatable observations. Without a theoretical interpretation, they are only based on Humean expectations and carry no causal implications. For the sound mind, illusions and dreams can be eliminated by their ephemeral character. On the other hand, we become aware of impressions from our mental activities. We feel emotions, communicate verbally, and erect complicated theoretical structures in science and society. Such processes derive from our subconscious apparatus, whose existence has to be part of the reality of the observer. The fundamental basis of this is not known; we do not know why we experience things the way we do, but instincts act on us, problems are solved, and relations are forged; we do not know how. The activity of the subconscious cannot be observed. Only its manifestations in consciousness can be experienced. Then it appears to be of equal character as the knowledge received about the external world.

I thus postulate a conscious entity, an archetypical observer, who receives information from the external world through sense impressions and from the internal psyche as mental experiences. In his mind, the two types of information are correlated and merge to form concepts: The objects of everyday experience call up ordinary words, results of experiments are interpreted in terms of physical theories, and physiological observations are related to psychic symptoms. We lack full knowledge about the deep sources of these observations. They may ultimately derive from some universal source beyond our present understanding. This eliminates all duality. The information entering the mind derives from mechanisms we can only try to depict from our limited knowledge.

However, to help us construct a world view, the conscious observer has access to some formal tools: This allows him to formulate symbolic representations of the experienced ambience. The child learns to navigate in an Euclidean three-dimensional space. We have an instinctive ability to learn languages for communication, and last but not least, the construction of mathematical complexes carries our symbolic manipulations far beyond the limits of our naive expectations. The existence of these built-in abilities must be taken to form part of the starting point for our inquiry.

The existence of the physical world and its experimental manifestations have led to the development of physical theories which, in spite of their abstractness, provide us with an amazingly efficient tool to describe and manipulate our environment. The symbolic machine has worked remarkably well here.

The theoretical understanding of our subconscious processes is less clear. Behaviorism is not sufficient; the mind insists on influencing us in unexpected and sometimes irrational ways. The sovereignty of the subconscious mind has been recognized by Carl Gustav Jung. His work is a systematic effort to find concepts describing the functioning of the common mind. This may ultimately also allow an understanding of aberrations now labelled as illness.

Investigating the workings of the psyche, Jung found that its subconscious material can be described by certain universal concepts which he named *Archetypes*. These are rational terms telling about a substructure common to all mankind. Its symbols carry meaning beyond the individual, it is the "collective unconscious". Its origin and mechanisms are hidden from us; we do not know to what extent it operates through our genetic heritage or forms in our interaction with the social surroundings. It, however, influences both our use of language and our actions in the world. It bears witness about the meaning of a world of mental entities just as real as the external world [16]:

But how do we assign meaning? From what source, in the last analysis, do we derive meanings? The forms we use for assigning meaning are historical categories that reach back into the mists of time - a fact we do not take sufficiently into account. Interpretations make use of certain linguistic matrices that are themselves derived from primordial images. From whatever side we approach the question, everywhere we find ourselves confronted with the history of language, with images and motifs that lead straight back to the primitive wonder world.

Thus the linguistic framework of humans plays an essential role in Jung's description of the inner realms. But his categories are not fixed and rigid as Kant's. They have formed dynamically during the evolution of man and they will adjust and reorganize with the extension of the sphere of human activity and knowledge. Like Feyerabend, Jung finds it difficult to fit the landscape of this inner world into the structure experienced as the external world. To start the discourse, less rational parts of the inherited language have to be used until more of the inner relations have been mapped [17]:

It is possible to describe this content in rational, scientific language, but in this way one entirely fails to express its living character. Therefore, in describing the living processes of the psyche, I deliberately and consciously give preference to a dramatic, mythological way of thinking and speaking, because this is not only more expressive, but also more exact than an abstract scientific terminology, which is wont to toy with the notion that its theoretic formulations may one fine day be resolved into algebraic equations.

Could it be possible that the myth of quantum theory was transformed into equations too early to admit our understanding to follow?

To conclude: *The conscious observer receives information from both an external and an internal reality. These enter consciousness from sources not directly available to the observer, but together they form the totality of experienced reality. On the basis of this information, the observer makes up theoretical descriptions using his capacity to introduce symbols and formal languages encompassing also abstract mathematical structures.*

2

TWILIGHT OF THE GODS

2.1. The dawn of darkness

My aim in this text is to treat two of the intellectual pioneers of the twentieth century: Ludwig Wittgenstein and Niels Bohr. The former has been influential in at least two major directions of philosophy and the latter stands for the currently accepted interpretation of the formalism of quantum theory. I want to compare them, spotlight similarities and differences, but foremost, I want to put them up as exemplars of intellectual originality and leadership. In their contexts, their lives will hopefully illuminate two trends of modern culture and society. The ultimate conclusions must be left to the reader; the issues are too complex to be condensed into precise statements.

The task set up seems to be impossible: Niels Bohr had no formal education in philosophy, and his writings on such topics are contorted and unattractive. The interpretations of his attitudes from the point of view of academic philosophy have been difficult and controversial. Likewise, Wittgenstein had a basic training only in classical physics. There is no indication in his writings of familiarity with modern physics nor any reports of verbal references to it. Thus this cannot be a book on philosophy or physics; there is not enough of factual material to satisfy either expert in these fields. It is only an attempt to see their activities as parts of a unitary whole in the evolution of human culture.

The text is written from my personal point of view; it is my reading of the textual material. As a physicist, I have some knowledge of the use and success of modern quantum theory. This gives me an advantage in reading Bohr, but it also poses a danger. As the so-called Copenhagen interpretation of the theory is not based on any consensus among its practitioners, the view of Bohr is not easily extracted. My knowledge of the various applications and the role of theory therein, may well confuse my judgement concerning Bohr's contributions. On the other hand, my insight into philosophy is based on extensive amateur reading, and the obvious danger is that I do not know enough of the literature, and what I know may be based on misunderstanding. When it comes to reading Wittgenstein, this may not entirely be a disadvantage. Not knowing most of the professional analyses of his contributions, may well keep my mind open to the task I have set up: to divine how much his approach to knowledge concurs with that of modern physics. As he himself used to claim, nobody is going to understand him. Thus I may be entitled to my own version of misunderstanding.

As personalities, Bohr and Wittgenstein were universes apart. The latter was immersed in a mystic feeling for the world, and he had a strong sense of personal moral responsibility in his life. He was acutely aware of his, partly hidden and forgotten, Jewish roots. The confessional religions seem to have had little impact on his attitudes. Towards fellow humans, Wittgenstein was short-tempered and sometimes positively rude. He was a rather unhappy person, who periodically lingered at the edge of committing suicide. Bohr, on the other hand, was pleasant and friendly towards colleagues and acquaintances. He grew up without any strong bonds to religion; possibly a reflection of the more liberal Nordic moral atmosphere as compared to the restricting Catholic milieu of Vienna. He lived in a lifelong happy marriage, and there are no indications of crises in his personal life. The drowning of his eldest son in 1934 caused grief but no lasting depression. He was periodically overworked and had to take a period of rest. This, however, was a natural consequence of his many commitments and huge amount of work.

In spite of their different religious upbringings, Bohr and Wittgenstein both stemmed from partly Jewish ancestry. I do not regard this as any cause for common mental attitude. It presumably only reflects the fact that the educated and creative part of the population in eighteenth-century Europe, with few exceptions, had Jewish relations.

Bohr and Wittgenstein do, however, share many features of similarity. Both came from wealthy families, where immersion in the culture of the day was characterizing the very atmosphere where they grew up. And this culture was that of the European continental scene when the nineteenth century turned into the twentieth. This differed from the cultural tradition in Great Britain and the still-emerging American cultural stage. Here I see the common roots of the method of approach used by both thinkers. New approaches were emerging in art, literature, psychology, science, and philosophy. The mood can best be characterized by the ill-defined word "existentialism". A revolution was going on, and the results are clearly seen in the world today. The human condition and its relation to authorities and freedom acquired new perspectives in this period. I find it significant that Bohr and Wittgenstein imbibed these trends from their childhood. The human individual is cast into the world, and it is, to a large extent, up to him to define his conception of this world.

Both Bohr and Wittgenstein put great emphasis on the role played by language in our approach to reality. They experienced a strong feeling of responsibility, when thoughts are transformed into texts. Both realized that intuitively divined insights about the world were not easily put down into words; in fact both despaired of an ultimate possibility to express anything essential in words, "Wovon man nicht sprechen kann, darüber muss man schweigen." As a consequence, both liked to work by oral communications. Some of the *Nachlass* by Wittgenstein is in the form of notes he dictated to collaborators, mostly younger ones but also to his senior colleague G. E. Moore (1873–1958). Bohr also is well known to have worked through his many articles by reading them to his younger disciples. They went through a multitude of versions, and the finally published articles bear clear traces of the difficult process leading to their birth. It seems that both men felt it easier to watch the line of their argument when an actual listener was present. It is also characteristic that both men exerted an immense influence on those who came in personal contact with them. This is not easy to comprehend when one bears in mind the public image of the two men.

Bohr published a lot during his lifetime, but this does not necessarily make it easy to discern his ideas clearly. Wittgenstein published only one book during his lifetime, the famous *Tractatus*, but he collected another manuscript into nearly publishable form. Both men left a voluminous collection of manuscripts and notes. This material has been subjected to extensive research. As far as I know, this is the first comprehensive effort to join the two manifestations of twentieth century culture into one concept.[1] As such it can only be incomplete and preliminary. But I am convinced of my opinion that they are but branches on the same tree. It lies in the nature of the issue, that I may not be able to present this conviction in a logically closed or scientifically compelling form.

2.2. Cultural backgrounds

2.2.1. Wittgenstein's Vienna

At the dawn of the twentieth century, the Habsburg double monarchy was something of a unicum [18]. Encompassing the lands of Hungary and Austria it stretched over many cultures and folks, which in itself guaranteed that it was full of internal tensions. The earlier Habsburg possessions in Spain and the Netherlands were no longer included, but the nationalistic movements of the nineteenth century had made the artificial entity an anachronism. Of its many languages, the Hungarian and Check tongues developed their unique identities taking over two of the three leading cities of the union: Budapest and Prague. The Balkan Slavic populations remained dispersed, which in time constituted the seed for the First World War.

Life, however, seemed to flow in its customary ways without any danger in sight. The dynasty was powerful but it persisted by ruling with moderation. There had been no great Habsburg heroes, the expansion had mainly occurred by politics and marriage. The people seemed satisfied with their rulers and their rulers with them.

When the artificial unity was shattered by the First World War, only fragments of the original greatness lingered on; in particular the center, Vienna, survived only as a relic of its former greatness. Those that had experienced the Habsburg Empire on the spot sorely missed it; nobody, however, seemed to want it back.

[1] An exception is the Master Thesis by Karri Sunnarborg entitled *Late-Wittgensteinian features in the philosophy of physics by Niels Bohr*. This interesting work submitted to the University of Helsinki in 1996 is unfortunately available only in Finnish.

Things changed slowly in Habsburg country; in many ways it appeared that time stood still. This was emphasized by the fact that the Emperor Franz Josef seemed to endure for ever. He became emperor of Austria in 1848 at the age of 18, brought to power by the revolutionary unrest of that time. He became king of Hungary in 1862. His very existence seemed to guarantee that nothing would change. He was to rule to 1916, and in that whole time all seemed to endure. The Austro-Hungarian double monarchy was unshakable. It was called *Kaiserlich und Königlich* which induced the author Robert Musil (1880–1942) [19] to dub it "Kakania", a word with childish fecal associations in most European languages. It was even more befitting as the Empire had no proper name of its own. Neither did it have a cultural unity.

This inhomogeneous entity was governed by relaxed moderation, which allowed the internal tensions to set without rupture. In his novel *The Man Without Qualities*, Musil depicts the situation as follows:

> A certain amount of luxury was employed, but far from the refinement of the French. A certain amount of sports and games did occur, but not with the frenzy of the Anglosaxons. Immense sums were directed to the military, but in any case exactly enough to certainly stay the second weakest major power. Even the capital was a little smaller than the largest towns in the world, but in any case considerably larger than the ordinary metropolis. And this nation was governed in an enlightened, utterly moderate and barely noticeable way by Europe's best bureaucracy, which only could be blamed for one thing: it considered as presumptuous and impertinent genius and ingenious initiative in private persons, who were not privileged to it by their noble birth or a commission from the state.

Adjustment seemed to be the order of the day, and all conflicts could be overcome by a relaxed attitude. All change was from the good towards the perfect. In May 1873 Vienna had experienced its Black Friday, and the value of stock depreciated badly. After that, however, the economy was stable, money that was saved appeared to be there for ever. An education or an official position once gained would prevail; when retiring the level of income was known and secure. The emperor himself was the guarantee for eternal permanence. Or so it seemed.

The world was changing, however. The comforts brought by improved technology were noticed, of course. But the infrastructure necessary was not the concern of the average citizen. The modern wind of trade and industry had reached Vienna. The new order required its pioneers. The traditional upper classes in Vienna devoted their working time, if such a concept existed, to the less arduous tasks of government and the military. Thus the challenge of modernity was taken up by the Jewish community. They formed a widespread but inhomogeneous part of the population, which was not cohered by their common religious culture. In fact, most of them had become secularized to the point where they were hardly aware of their roots. But they had the knowledge, the contacts, and the linguistic skills to organize the Industrial Revolution. Wittgenstein's family belonged to that group of successful entrepreneurs. His father's involvement in the countries' industrialization made him one of the richest men in Europe.

Here, however, we encounter one of the great paradoxes of the Jewish inheritance. It seems to give a drive and predisposition for success which goes far beyond the influence of ethnic society and religious tradition. Nowhere has this been so remarkable as in *fin-de-siècle* Vienna. Jews played a central role in the transition to the new technological society, but they also took the lead in the cultural revolution signalling the dawn of the twentieth century. In spite of all worldly success, they never seemed to loose their urge towards culture and learning. They formed the cutting edge in all novelty of the petrified Habsburg Empire. This dominating role may have cost them dearly in the new political situation emerging in Germany.

But even in Vienna, the Jews had to remain in the shadow of the Austrians. The emergence of the various ethnic identities was paralleled by the growth of pride in things Germanic. The edge of nationalism cuts two ways. The novelist Stefan Zweig tells that his father never dined at the fashionable restaurant Sacher [20]. Not for economic reasons, he had amassed a considerable fortune in the growing textile industry. No, at Sacher he would meet the upper-class members of the Habsburg bureaucracy. However much he felt Vienna to be his home, among them he did not feel comfortable.

Under the surface there were undoubtedly strong forces directed against the Jews. Freud tells the story he had from his father: he had been walking in his home city of Freiberg dressed in his best Saturday clothing including a fur hat. A Christian throws the hat into the dirt of the street and commands: "Down from the sidewalk you Jew!" Freud asks what the father did, but is told that the father only picked up the hat without a reaction. No wonder that Freud later despaired about achieving the position of professor in Vienna.

The political situation in the Empire was tense. The centralized leadership was strongly imposed by the emperor's bureaucracy and tight censorship. Under the surface, however, new and threatening ideas were emerging. Liberal concepts and socialist considerations were coming up, moving towards the form of political organization. Inner tension was growing; not only the ethnic separatists but also the deprived part of the population were gaining influence. The political movement also led to a strengthening of the German character of the Empire, which tended to separate out the Jewish population which had seemed so well integrated. As a consequence Zionistic ideals emerged. The drive was provided by an original Budapest snob Theodor Herzl, who became convinced that only an Exodus would solve the problem posed by the Jewish situation in Vienna. His initiative grew to a politically significant influence in the Empire and eventually in the world.

Thus we have the bizarre situation that both the roots of the Holocaust and the Exodus sprouted from the same soil. Such was the paradoxical situation in the *fin-de-siècle* Habsburg Empire.

But as we have noted, culture played a central role in the daily life of the Viennese. The wealthy industrialists both sponsored various arts and partook in their execution themselves. After the economic conciliation of the third quarter of the nineteenth century, the young generations found less pressure to deliver new results and spent more of their time in leisure. The financial crisis of 1873 had shuffled the roles somewhat, but the cultural scene had not changed appreciably. The bourgeois devoted a considerable part of their time to culture irrespective whether they were affluent or only coping. The interests were broad and the commitment was intense.

This is the time of the emergence of the café culture of Vienna [21]. The participants spent prolonged hours in the café houses, sipping a single cup of café or a glass of wine, but there having access to major newspapers of the time. There they also met and debated hotly a multifarious spectrum of cultural issues. Each one had broad interests and the exchange of ideas had a high priority.

The café house culture was promoted by the fact that housing was scarce in Vienna. In the half-century before the First World War, its population grew by a factor of four without a corresponding increase in the availability of housing. Thus the only meeting places possible were the public cafés. This, of course, did not concern the affluent ones, but interested in the discourse as they were, they also joined the intellectual crowd.

In spite of the seeming stability, the active-discussion participants must have felt that the times were approaching a cataclysm. No other explanation can account for the many consorted breaks with tradition that occurred and all the novel conceptions that emerged:

Vienna was immersed in music: Strauss and Lehár brought the classical tradition to the level of everybody, and the waltz became nearly the trademark of Vienna. But out of this emerges Arnold Schönberg, who with his atonal music utterly modernized composition.

In architecture the heavy imperial style was transformed into a more functional modern design by Adolf Loos. In painting the artist Gustav Klimt introduced the strangely abstract decoratism of art nouveau, whereas Oscar Kokoschka struggled to capture the Viennese ambiguity in pictures lacking any close contact to realism. Klimt initiated the Viennese art direction called the Secession, which has become an integrated feature of Austrian twentieth-century art. The building they erected to display their art was financed by Wittgenstein's father. In parallel with the Norwegian Edvard Munch, the Austrian pioneers were to initiate the art movement of Expressionism.

The psychological atmosphere in Vienna was tense and complicated. Stefan Zweig points out the hypocrisy characterizing the sexual situation in the city. He states the case [20]:

> If it is not feasible to do away with sexuality, then at least it must not be visible in the world of morality. A silent pact was therefore reached, by which the entire bothersome affair was not mentioned in school, in the family, or in public, and everything which brought its existence to mind was suppressed.

It is easy for us, who since the time of Freud know that whoever seeks to suppress the consciousness of natural desires not only fails to remove them but dangerously displaces them into the subconscious, to laugh at the unenlightenment of the naïve technique of concealment.

Women were supposed to remain chaste, and men married late. As a consequence, prostitution abounded and official condemnation hid the widespread acceptance of the situation. Among the bourgeoisie, marriages were often arranged and men consequently felt little love for their partners. Married women were thus subjected to the pressures of extramarital affairs, and their lives vacillated between desire to yield and frustration at abstinence. The situation is well depicted by the famous authors of the time: Hugo von Hoffmannsthal (1874–1916) and Arthur Schnitzler (1862–1931). These provide the perfect illustrations of the sexual situation and its Freudian developments. Also the short stories of Zweig are imbued with the social tension of the society. Even earlier, the author Frank Wedekind, in 1891, had presented a horrible view of puberty in his play *Frühlings Erwachen.* As a curiosity may be mentioned that he lived some time with Strindberg's ex-wife Frida Uhl.

The taboos surrounding the sexual situation still lingered on, but a novel attitude was gradually paralleling its emergence in art and literature. The psychiatrist Richard von Kraft-Ebing was bringing out into the open the sexual needs and pathologies of both men and women. He was widely travelled, and had studied in Heidelberg, Zurich, and Berlin. In 1870 he was appointed professor in Strasbourg, then Prussian territory. Being the foremost expert in his field, he published the book *Psychopathia Sexualis* in 1886. This mapped the sexual habits of humans in a way which has been surpassed only in our postwar times. The book took a pioneering attitude of tolerance towards the varieties of sexual behaviors in humans. However, in the preface of the first edition, the author states [22]:

> The scientific study of the psychopathology of sexual life necessarily deals with the miseries of man and the dark side of his existence, the shadow of which contorts the sublime image of the deity into horrid caricatures, and leads astray aestheticism and morality.

The frankness of the book was partly offset by the fact that the case histories were originally included in Latin only. The work, however, became the standard text in the field and ran to a multitude of editions. The stage was set for Freud and the revolution in psychology.

This book was certainly known to Sigmund Freud, when he started his extensive clarification of the physic factors behind mental problems. Kraft-Ebing first presented his theories as a professor in Vienna. He failed originally to be convinced by Freud. When Freud in 1896 presented his theories, Kraft-Ebing labelled them "A scientific fairy tale". Later, however, he supported Freud, who acknowledged this in his *Theory of Sex* in 1905. By that time, Kraft-Ebing was dead and we do not know how the collaboration would have developed.

Moral values were undergoing a rapid development which was mirrored in the literature and art of the time. However, there is no doubt that the Viennese situation was a major influence in the development of Freud's ideas. He assigned a central role to the sexual experiences of the individual, and he constructed the edifice of his theory on them. Freud was a good empiricist; it is fully conceivable that his observations on his female clientele motivated his view. In *fin-de-siècle* Vienna,the social situation supports this assumption. Later, our conception of the sexual realm has changed greatly; the theories of Freud have been superseded and expanded, but he, no doubt, initiated the movement of "depth psychology" which forms part of our present view of the human condition.

The Viennese felt lost in their chaotic culture, and their hectic lives were conducted under a dread seemingly without origin and form. The situation was characterized by the witticism: "The situation is hopeless but not serious." Nobody embodied this situation more than the writer and journalist Karl Kraus, who became the prophet as well as the scourge of the atmosphere of *fin-de-siècle* Vienna.

Kraus was born in 1875 and started to edit and publish the fortnightly journal *Die Fackel* in 1899. It was to appear until his death in 1936 and became one of the central organs for mirroring the moods and events in the Habsburg Empire even after its demise. The journal cultivated a meticulously refined satirical style; the language was ingeniously utilized and the targets were events and persons easily recognized by the eager public. The aim of the anti-paper was to bring out all the corruption and hypocrisy behind the idyllic facade of the life in Vienna and its Empire. The victims were the superficial joys of the popular operettas, the corruption of the authorities, the idealism of Zionism, and eventually the slaughter and the futility of the First World War.

A special target for the satire was the leading Viennese newspaper *Die Neue Freie Presse*. This was devoted to journalism at the highest professional level. Its style and opinions were respected far beyond the borders of its home range. For Kraus, the very excellence of the paper stood in sharp contrast to its lack of objectivity. The strict censorship of the Empire made it an instrument for the powers that be. The strong capitalistic influence on the bourgeois classes of Vienna, made the leading paper serve the purposes of the industrial interests. Finally, it officially condemned the widespread prostitution, but it had no qualms to profusely publish its advertisements. All in all, *Die Fackel* kept the public alert to the overt hypocrisy of the society.

Like so many Jews of the Empire, the attitude of Kraus towards Jews could be labelled as self-hatred. This complicated issue impregnated the social atmosphere of Vienna, and left its traces on the pages of *Die Fackel*. In the same manner, the view of women was condescending, bordering on contempt. The latter feature was amplified by the writer Otto Weininger, who published his controversial work *Geschlecht und Charakter* in 1903 and promptly committed suicide; he was only 23 years old. We have reasons to return to Weininger later.

Kraus was the dominating force behind *Die Fackel.* Originally he was supported by some of the leading authors of Vienna, but after 1911 he wrote everything himself; a single exception is a column by August Strindberg in 1912. Its witty style and urgent mission was, however, marred by its attitude to Jews and women. This politically incorrect approach appears repulsive to modern readers, but it may not have given the same impression to the public of the time.

The critical attitude of the times towards insincerity and pretence ignited an interest in the ideas of the existentialist Danish writer Søren Kierkegaard (1813–55), who emphasized the total commitment needed in human life. In a less philosophical manner the Russian author Leo Tolstoy was admired for his ethical attitudes to life and art. Kierkegaard was entering the cultural circles of Austria through his advocate Theodor Haecker, who translated his works into German and wrote several books about him. Haecker realized that the author was skeptical towards conventional formulations of the fundamental issues of human life. Kraus was seen as a latter-day incarnation of Kierkegaard. Like Kraus, Kierkegaard had come to realize that such questions could not be cleared up by the use of language; only undoubting faith can overcame the anxiety central to existence. Moral truth cannot be communicated, it has to be experienced. Such was the image of Kierkegaard that entered the discussions in Vienna. It mirrored many facets of the mood of the period, and one may argue that the introduction of Kierkegaard was an influential factor in the later developments of European existentialism.

In the *fin-de-siècle* Empire, old traditions conflicted intensely with emerging novelty in all areas of human life. An eerie feeling of impending doom was felt in many circles towards the twilight time of the Habsburgs. The situation in Vienna was steady but not stable. The hectic cultural life was racing towards its ultimate end; what that end would be was beyond surmise. The First World War offered an end, anticipated but not foreseen. After the war nothing was to be as before, and the need for a new future was obvious.

More sinister than the glossy life of Vienna, the spectre of doom lingered over the peripheral parts of the Empire. Prague was the home of the author Franz Kafka (1883–1924) [23], who describes society in the agoraphobic vision of the *Trial*; its action and conclusions forever escaping the poor K. It is not clear to what extent the empirical bureaucracy provided the impetus for this feeling of getting lost in the social machine.

The streets of Prague were known as the scene where the ghastly Golem roamed. Created by Rabbi Löv, according to myth, this artificial being spread fear and havoc over the city, like so many successors in today's movies of horror. The Golem rose again as the creation of Frankenstein, who again showed the hazards of reckless playing with the forces of Nature. Man's failure to control his creation opens the question: can man control himself?

Combined with the Kafkan vision of a social machine without compassion and mercy, the theater of terror offered a new stage, a future society governed by technical networks of beings beyond empathy and pity. The author Karel Čapek (1890–1938) presented this horrifying vision, and created at the same time the name and the concept of a "robot"; a machine not only employable in technical tasks but also suggesting the image of a technical Golem, powerful but eventually beyond human control. Does not the spectre of the robots derive its power from the *fin-de-siècle* Vienna?

Still further away, in the mountain ranges of Transylvania, ruled Count Dracula; allegedly spreading vampires over all Europe. Still another threat was lingering over a society without future and without hope. The atmosphere derives from the twilight times of the Habsburg Empire. The Empire departed, but the horrors it had created still stay with us, become ever more global, and remain the underpinning of our unspoken fears and misgivings even today.

2.2.2. Bohr's Copenhagen

Something rotten in the state of Denmark

Denmark had been a major power in northern Europe, and it had acquired a status of wealth by global trade. The country was in close cultural contact with the Germanic world. It was in good contact with the literature and philosophy of the Continent, and its educational level was advanced and high. It experienced a strong blow to its economic strength when the English fleet attacked and conquered Copenhagen. This was caused by the alliance of Denmark with Napoleon's France, which made the British include the country in their blockade of the Continental ports. Following this the country suffered an increased inflation, which in 1813 caused the bankruptcy of the national bank. In 1814 Denmark was forced to hand over Norway to Sweden.

Towards the end of the nineteenth century, Denmark seemed to be a happy and advanced democratic country. However, it had lost its greatness and was adjusting to its new role as a minor power in Europe. On February 1, 1864 Bismarck had ordered his troops into the regions of Schleswig and Holstein, which up to that time had been under Danish rule. The region was considered part of German territory, the population was partly German-speaking, and the access to the North Sea was attractive to the expansive German politics. The Danes defended their domains, but the outcome was a given: Germany took over one-third of the kingdom of Denmark and one-third of its population. The loss resulted in a period of depression and paralysis, which may be paralleled in the gloom of the Kierkegaard philosophy. However, the country recollected itself, and its inner evolution compensated to some extent for the loss of status in the world. A period of cultural development followed; literature and music flowered and the national self-esteem rose anew.

The intellectual atmosphere in Copenhagen was still strongly under the influence of Søren Kierkegaard, who had died already in 1855. The gloomy world view of this thinker, who taught that rational arguments must stop and be replaced by discontinuous leaps of faith, formed a basis for a re-evaluation of all values later advocated by Nietzsche and taken up by the existentialist movement. Niels Bohr had read Kierkegaard, he admired his use of the language but did not accept his arguments and ideas. Thus he should not have been influenced by his thinking. It is, on the other hand, possible that the tone of the philosophical arguments he encountered during the discussions in his boyhood left unconscious traces on his way of thinking. I will return to this issue later.

During the Copenhagen student debates in 1824 a philosopher and author Poul Martin Møller had presented parts of his novel *A Danish Student's Tale*, which contributed to the debate gaining more and more ground in the city. The student debates also offered the first forum for the emerging thinker Kierkegaard. Thus a circle is closing. Møller was a colorful person, who was appointed professor of philosophy first in Christiania (now Oslo) and from 1831 in Copenhagen. This left his novel unfinished. As a professor Møller lectured on Hegel and the history of philosophy. These lectures were attended by Kierkegaard, who acquired an extensive knowledge of antiquity from them. As a teacher Møller was more known for his aphoristic presentation than for discursive arguments. Møller died in 1838.

Møller's novel was read and liked by Bohr. He there found a description of the inactivity which may derive from eternal contemplation of the activity to be undertaken. Thus thinking about writing becomes an obstacle to actually starting the activity. Bohr highly liked this tale and quoted it often.

This may be seen as the story of a personal development, how the individual finds his way and task in the emptiness of the universe here symbolized by contemporary society. In this tale, we encounter the Licentiate, who, in spite of the best intentions, is unable to get his work under way [24]. He is challenged by the visiting relative.

Nothing which can be done should be postponed to tomorrow.

Licentiate: If you deny me that, you must necessarily allow me that it just as well may be done one minute or a second later, and this cursed possibility lasts as long as one breezes. Nobody can show me how such a postponement which lasts a second or a fraction thereof should have any disastrous consequences, and because nobody can show me that, my life goes into a standstill.

My unending consideration of this implies that I do nothing. Furthermore I come to think of my thoughts about it, indeed I think about my thinking thereof and I devote myself into an eternally descending line of "I's" who observe each other. I do not know at which "I", I may stop, and in the moment I stop at one, there is again an "I" which does the stopping. I become confused and feel dizzy as if I looked into a bottomless abyss. The considerations end by giving me an annoying head ache.

This sentiment was appreciated by Bohr, who considered it an impressive demonstration of the dilemma of an acting subject in the world of doubtful grasping of reality. This has fixed a position for the Danish student in the history of quantum physics. Its author Poul Møller has achieved a certain fame as one of the few known individuals influencing Bohr.

This fame is, however, to some extent ill deserved. It seems to derive from an excerpt in Goethe's work *Dichtung und Wahrheit* usually translated as *Fiction and Truth*, part I, book 7:

I had often pressed my friend Behrisch, too, that he would make plain to me what was meant by experience? But, because he was full of nonsense, he put me off with fair words from one day to another, and at last, after great preparations, disclosed to me, that true experience was properly when one experiences how an experienced one must experience in experiencing his experience.

As this work appeared in 1811, it is likely that it was known to Møller when he wrote his novel. This is difficult to prove, but acknowledging the central position of Goethe, it seems likely.

2.2.3. Kierkegaard as symbol of the times

For a long time the European scene was dominated by stability. Political matters were prescribed by the rule of the Estates and spiritual matters were taken care of by the Church. The dominance of Catholicism was broken by the Protestant Reformation, but still the clergy stated what was the official opinion and what was not.

The ideology was, however, to yield to new views. Behind the upheaval of the French Revolution lay the swell of liberal ideas emerging during the eighteenth century. It seemed that suddenly everything could be re-evaluated; religious doctrines could be overcome and the nobility could be eliminated. The world was ready to redefine its basic tenets. The new freedom was intoxicating, but it left the humans without signposts to direct their lives.

The grand political system envisaged by Hegel may be seen as an attempt to bring order into a universe without a firm foundation. It was a good try, and it seemed to be satisfactory for a time. In the form of Marxism, it indirectly showed the way to a new systematic certainty whose footprints are manifestly visible today.

But Hegel's system was not to show the road to our modern world. After an initial fascination, its shortcomings became obvious and new roads had to be found. Humankind had to define itself in the novel universe without solid ground to support the mental constructions, without self-evident dogma and a governing God. The duty to show the way into this process became the fate of a lonely Danish writer and thinker, Søren Aabye Kierkegaard. In spite of writing all his texts in his native Danish he was to become a European celebrity and virtually the father of the later existentialist movement.

Different countries reacted differently to the religious feeling of being lost. In Denmark, the mighty personality of Nicolai Frederik Severin Grundtvig (1783–1872) dominated the scene. He started as a scholar devoted to old Nordic literature, but after a religious awakening he proceeded to vitalize both the Danish church and the religious life in his country. His strong polemics created opposition, and he came to develop his ideology of a free church in the liberal atmosphere in England. Grundtvig was later able to return to Denmark and became something of a father figure for a generation of younger theologians. Their opposition to an old-fashioned and rigid view of religion served to keep the debate open for a long period. Matters of religion were certainly clearly visible to the intellectually alert in Copenhagen during the early nineteenth century.

Søren's father, Michael Pedersen originated from a simple country background. As a young boy, the father got the opportunity to move to Copenhagen, where he eventually entered the cloth and textile business and managed to create a solid middle-class position for himself. In the economic crash of Denmark in 1813, Michael Pedersen had managed to move his assets into bonds that retained their value. Consequently, he did not suffer and emerged comparably wealthy. He had seven children, with Søren being the youngest; of the four boys, only he and his elder brother Peter Christian, who was eight years older, lived to mature age. Niels Andreas, who was four years 'older', died in the USA at the age of 24. Even one of the sisters died prematurely at the age of 25.

Michael Pedersen was not one to easily adopt novel and liberal ideas. He gave his boys a stern religious upbringing which came to dominate their lives and views. He, apparently, suffered from a melancholy state of mind combined with a conviction of moral shortcomings. This was not a healthy atmosphere to grow up in. The elder boy, Peter, traveled in Europe, imbibed the cultural mood of the time, returned with a German doctorate, and took up a career in the Church. Søren was less ambitious, at least in the beginning. He was supposed to study theology but apparently concentrated his efforts on philosophical and esthetical topics at the university. He claimed that his father's religious indoctrination had made his life miserable [25] and [26].

Choosing an attitude to life

Søren Kierkegaard certainly felt the burden of moral and religious duties as urgent and heavy. He experienced a strong feeling of moral obligation combined with a despair lest he be unable to live up to the requirements set by his convictions. In this aspect he seems to be very similar to Ludwig Wittgenstein who also felt isolated and left on his own in sorting out the duties and requirements of life.

In contrast to his brother Peter, Søren did not pursue his studies with any efficiency. He took part in happy student life, involved himself in the debates of the time, and hid much of his inner doubts about the meaningfulness of it all. He possessed a strong satirical talent and managed to hide much of his deeply felt seriousness behind a verbal camouflage giving him a reputation as a fearsome opponent and a master at arguments putting down pompous and superficial pretensions. We do, however, know much about his inner considerations, because he was a prolific writer of notebooks expressing the attitudes and queries of his inner life.

Søren's doubts directed him to seek kindred spirits in literature. Faust was an obvious one: at the time of course the works of Goethe were everywhere present. Another character who fascinated him was Ahasverus, the wandering Jew, whose fate had come to symbolize the human quest for the meaning of life after the collective fall into sin against living God. Another Dane, H. C. Andersen, struggled in vain to write a play about the theme of Ahasverus.

A most curious hero of Søren Kierkegaard was the German folklore character of Till Eulenspiegel. He was a typical trickster figure, who vacillated between being the ingenious thief and the innovative prankster. The stories are often crude and based on bawdy verbal humour. Kierkegaard's fashination with this figure serves to prove his strong attraction to the humorous and satirical. Interpreted at a higher plane, Eulenspiegel may be taken to symbolize the outsider, who is not only a social outcast but also the superior individual rising above good and evil. He has taken full responsibility for his life and chooses freely the course of his actions which also implies full responsibility for the results.

In the same vein, Kierkegaard feels attracted to the Don Juan figure as put forward by Mozart. In his later writings, Kierkegaard extols Don Juan as a person above considerations limited by reflection and language. Here again, we encounter the individual making his own rules and finally having to face the ultimate fate of all outsiders, to be condemned to his personal version of hell. Interestingly enough, one of the later leaders of the existential philosophy, Albert Camus, chooses Don Juan as an example of a person committed to a full life in a world deprived of divine direction.

An influential teacher

A person with lasting influence on both Kierkegaard and the Danish cultural scene was the author and philosopher Poul Møller. He was born in 1794 and lived only to 44 but he had a diverse and influential career. During the years 1826 to 1830 he was professor of philosophy in Christiania (now Oslo); from 1831 he was professor in Copenhagen.

In order to live up to his duties as a teacher Møller abandoned his literary ambitions and concentrated on philosophy, specializing in the topical issue of Hegel. In addition to Hegelian thought, Møller had a deep interest in classical culture, predominantly Greek, which gave Søren Kierkegaard a good basis for his departure from the conventional wisdom. Møller's own contributions to philosophy were not systematic and rigorous; he was a master of short aphorisms, which also characterize his posthumously published works *Scattered Thoughts* (1839–43).

Møller's reputation does, however, rest mainly on his literary production. Foremost here is the unfinished novel *A Danish Student's Tale*. This may be seen as the story of a personal development, how the individual finds his way and task in the emptiness of the universe here symbolized by contemporary society. As described above, we encounter the Licentiate, who, in spite of the best intentions, is unable to get his work under way. This lays down the mood; the human individual feels unable to choose an action, and hence nothing is achieved. There is no doubt that Møller here captures the mood of the existentialist message of the twentieth century. According to F. Coplestone [27]:

> Existentialism lays emphasis on the human situation or condition. We are told, for example, that man finds himself in the world, that he is a being in the world. We are told that he is a finite, unstable being, menaced by death from the start. We are told that he is free, that he transcends his past and inevitably shapes himself by his free choices in such a way that he is never a mere "object" until death extinguishes his possibilities. As a conscious free being, man stands out from the background of nature.

This echoes the dilemma of the Danish Licentiate at the beginning of the nineteenth century. No wonder the legacy of Kierkegaard's lingers over present-day existentialism. Also, as we will see, the writing of Møller has influenced the thinking of modern physics through Niels Bohr in, perhaps, unexpected ways.

The difficulty to find the right moment for action goes a long way back. The argument was put forward by Aristotle as an objection to a finite beginning of the universe. If nothing existed, there could be no cause singling out a definite beginning from the unstructured nothingness before creation. If there could be no reason to start the universe at a specific time, and the universe exists, there can be no beginning. This argument may well have been known to the classically knowledgeable Møller. Later it was to pose a problem to the Christian doctrine of divine creation. The issue was taken up by Augustine, who claimed that if time was created at the same instant as the universe no conflict arises. Curiously enough, it seems that modern cosmology solves the problem in the same manner.

Finally a degree

In the summer of 1838 Søren's father, Michael Pedersen, died at the age of 82. It seems that Søren Kierkegaard finally was freed of the inaction imposed by the dominance of his father. He suddenly took up his studies and conducted them to an end. A contributing factor may well have been the death of his mentor Møller, but it seems likely that a factor was the difficulty in combining Søren's attitudes with the religious zeal of his fathers.

Søren's enthusiasm to study surprised many; he had become the archetype of the eternal student. However, in 1840, he passed the written and oral examinations for his university degree in theology. His performance was not outstanding, but he had carried it to completion in record time. Kierkegaard, however, still had to write and defend a thesis.

Kierkegaard got a dispensation to write his thesis in Danish instead of the customary Latin. He chose the title *On the Concept of Irony with Constant Reference to Socrates*. Here he was able to combine two of his interests: the ironic view of discourse combined with his knowledge of the classics. The whole work was very much in line with his track record and abilities. Irony was his trademark and he could excellently apply it to academic writing. In fact, it has been claimed that the whole presentation was an example of metalogics; the subject of the work, "irony in writing", is not argued for but shown by the presentation of the thesis as an example. If this is so, it seems to connect the work to Wittgenstein's claim that the profound things of human life cannot be stated but only shown.

Kierkegaard's dissertation did not meet with unequivocal acceptance, but in October 1841 he was awarded the title of *magister artium.* He now had a degree, and in front of him lay the decision what to do with the rest of his life. He had initiated his career as an author by publishing the booklet *From the Papers of One Still Living* in 1838. He could fulfill this promise by remaining a free writer; the inheritance after his father formed the basis for this. On the other hand, the financial situation did not allow for the support of a family. In that case he would have to apply for a position, presumably in the Church. Then, the duties and surroundings might well hamper the writing. Thus a decision needed to be taken.

Kierkegaard did have a candidate for the marriage plan; namely the girl Regine Olsen whom he met in 1837 when she was only 14 years old. After finishing his studies, he felt economically strong enough to propose, and the couple entered a formal state of engagement. They exchanged numerous letters, but the relationship did not seem to progress. Eventually, however, Kierkegaard had second thoughts and broke off the engagement. Somewhat unexpectedly, Regine tried to oppose the break, but to no avail. Kierkegaard could not fit her into the plan of his life. Still, when she got engaged and married somewhat later, he was disappointed at the outcome. Somehow he had seen the story as a love of a lifetime, and he expected her to live on the memories as he was to do.

An interesting parallel to Kierkegaard's love affair is presented by the later exponent of existentialist attitudes, the author Franz Kafka (1883–1924). In spite of two engagements, he never could decide to marry. His contemplation about the difficulties is found in his diary of 1913:

> I have to be alone a lot. Whatever I have accomplished is only a result of being alone....
>
> I was often a completely different person in front of my sisters—this was especially the case earlier—than in front of other people. Fearless, exposed, powerful, surprising, moved as otherwise only through writing. If only I could be so above all through my wife! But would this not then deprive me of writing? Of all things not that, not that!

Here we again echo the truly existentialist author's fear of becoming dependent. Only the lonely is free to follow his destiny.

Literary career

When Kierkegaard finally devoted himself entirely to writing, his output became prolific. In a few years he published what was to establish him as a leading author of the time and a classic of the future. The final form of his output owed a great deal to his mental development relating to Regine, but the rapid writing indicates that much of the material was well prepared during a long time.

The main works of this period: *Either-Or* (1843), *Repetition* (1843), *Fear and Trembling* (1843), (1844), *The Concept of Anxiety* (1844), *Stages on Life's*

Way (1845) established a reputation. Familiar both with the Hegelian style of argumentation and the dialectic method of Socrates, Kierkegaard chose a presentation where contrasting points of view alternated and broke against each other. In order to achieve this, he introduced various pseudonyms into the discourse, where he could adopt different attitudes and concomitantly present contradictory aspects of a variety of issues. In this way Kierkegaard was able to avoid the fate of the Danish Licentiate to let the various "I's" prevent each other from acting. Allowing the inner split of his mind, Kierkegaard could let them all present their views. To discern what was the "real Kierkegaard" in all this is difficult; in the end it may have been his intention to make it impossible.

The magnitude of the work is immense, and both the linguistic and the philosophical presentations are hard to penetrate. It is, consequently, remarkable that the works became popular and their author a celebrity. This, perhaps, says more about the intellectual atmosphere of Copenhagen at the beginning of the nineteenth century than it says about Kierkegaard; in a less educated milieu, he may not have been noticed at all.

Either-Or took up the two issues dominating Kierkegaard's mind: the ironic detachment from reality and the necessity to live up to the ethical requirements of human existence. The former echoes the topics of his thesis, the latter his anguish about the relation to marriage. Together the two parts may be seen as a Hegelian thesis–antithesis tension. The synthesis was still lacking.

Kierkegaard's next two big books take up the moral duty of a human again. As a starting point he chooses Abraham, who is challenged by God to sacrifice his son. This absolute imperative Kierkegaard takes as the starting point in arguing that the human has to posit himself on the multifarious issues confronting him in life. He was still much dominated by his Christian upbringing, but he was not yet ready to shoulder the uncompromising obedience imposed by God on Abraham. In his argument for radical decisions, he appears to justify his action in the affair with Regine. He expected the public to react with dismay, but to his surprise the works became a success.

However, the quest for more of a synthesis goes on. Kierkegaard's concept of "Anxiety" was developed from a long tradition. The German Romantic movement had been dominated by a feeling of "*Weldschmertz*", which could also be translated as "melancholy", which is less drastic than the earlier translation "dread". Modern psychologizing parlance has made it hard to grasp the exact feeling conveyed by Kierkegaard. The popular word "alienation" captures one aspect of the feeling.

In 1844 Kierkegaard published two works. These have a more academic tone than the previous texts. However, in the spirit of all his presentations, the arguments are dialectic containing contradictory and obscure statements. The general movement of the discourse is to strive towards a synthesis directing the choices implied in the human life.

In the work *The Concept of Anxiety*, Kierkegaard considers the dual parts of human life, body and mind; these can only coexist by the existence of the spirit.

But he finds that no rational argument, neither Socrates' dialectics nor Hegel's "logic", can lead man to a rational decision about his being. The choice is free, and Kierkegaard experiences anxiety when confronted with the choice: Anxiety can be compared with dizziness. The person whose eyes happen to look down into the abyss becomes dizzy. Here we are reminded of the Danish Licentiate and his fatal state.

Kierkegaard is immersed in a state where the strong experience of sin pervades his thinking. Faced with the insight that no science, including psychology, can direct man's choice of fate, Kierkegaard concludes: "Then everything is changed, and when freedom gets up again it sees that it is guilty. Between these instants lies the leap, which no science has explained or can explain." The individual faced with his situation in the world has to make his own decision about what to do with life. If there is a God to be reckoned with, he does not prove himself, but his presence is manifest to influence the choices. No wonder the existentialists of the twentieth century were to emerge from Kierkegaard's position.

Much of the work *The Concept of Anxiety* is written in the manner of Hegel. There may be an irony in this, because the superficially logical development of the ideas end up in a surrendering conclusion that no answer can be found along this path. Here an analogy with Wittgenstein's *Tractatus* suggests itself; the ladder of arguments leads just that far, then it has to be abandoned for what it cannot encompass.

In the book *Stages on Life's Way* Kierkegaard suggests a synthesis of his considerations. A road emerges, which directs the individual from a theoretical to a spiritual attitude. The starting point is an esthetical attitude. In spite of Kierkegaard's interests in science as a young man, his mature activity did not touch upon the arguments of "exact science". The personal situation in life directs the individual to make moral choices. This leads him to an ethical position, where personal responsibility emerges. However, there is no finality in this, and the ultimate position is termed a religious one. Progress on this road cannot be directed logically; somewhere a leap of faith or a decisive decision has to be taken.

A "grand" conclusion?

It seemed that by 1845 Kierkegaard had reached some kind of completing attitude. He still had much to say; in the work *Concluding Unscientific Postscript* (1846) he added six hundred pages of further discourse [28].

The status of Christianity was still the central issue. After having settled with the rational road to religion, Kierkegaard now took up the historical roots of Christian faith. The accidental truth of history can never serve as the proof of a truth of reason. Here Kierkegaard chooses Gotthold Ephraim Lessing (1729–81) as his spokesman. This may well be a pose adopted by Kierkegaard. Lessing was a Romantic and mainly known as an author. To select him as a vehicle, Kierkegaard could remain outside the circles of academic philosophy and choose a point of view opposite to the strictly academic Hegelian one. Lessing also may have appealed to Kierkegaard's attraction to the ironic.

From Lessing derives the concept of a "leap". When rational arguments fail, a discontinuous transition to a new attitude becomes necessary; when an eternal truth is stamped on a historical fact, this can only be performed through a leap. The transition is a discrete one, you either decide to make it or not; one cannot be in the process of making it. The choice when one becomes a personality is to become totally subjective.

The work analyses the concept of "truth", which takes on a strongly personal flavor. To be true is to be true to one's personality. The argument is that no reality has been attributed to a phenomenon before it has been transformed into a thought. But, because truth is also a thought, it cannot be arrived at by contemplating the external world only.

As usual Kierkegaard is making many of his claims by attacking them. To extract a logically coherent position is difficult, but this is an essential part of what he wants to convey. Only by abandoning the rational approach and taking the leap, can we arrive at the truth. The only aid we may enrol along the way is irony. This makes it even harder to extract the serious parts from Kierkegaard's writing, but this is presumably part of the argument too. Already Socrates knew how to argue through irony.

An ironic twist in the book *Concluding Unscientific Postscript* is the concluding remark by one of Kierkegaard's more popular alter egos, Johannes Climacos: Here we meet the statement that the entire book is superfluous. It is not supposed to induce by any means any real person to be the reader. Taken at face value, this seems to revoke the whole work. However, the emphasis is on the words "real person" and "reader". Maybe Kierkegaard distinguishes between the reader who has to benefit from the discourse and the real person that already has taken the leap and cannot benefit from any further reading. This interpretation reminds us of the attitude taken by Wittgenstein in claiming that his *Tractatus* can only be understood by someone already having had the same thoughts and then no further argumentation is needed; the ladder of rationality can be thrown away.

There is no doubt that Kierkegaard regarded his argumentation to be final at this point. Like Goethe's Faust he had reached the ultimate destination and consummated this by a huge manifesto just like Goethe. The play was finished, the author had said what there was to be said, and the rest was for posterity. Kierkegaard indeed seems to have expected to leave the world at this time. But he was to live another ten years.

At this point we may draw an analogy between the dialectic method of Kierkegaard and physics. An idea in science emerges after accumulation of observations in nature. At some point the unexplainable insight appears, a researcher formulates a theory, which can be tested and finally be accepted. We have to accept as a fact that the empirical evidence is never sufficient to imply the law by logical argument. However, when grasped and tested the law becomes "The Law" and takes on a life of its own. It may be amended in time, its range of validity may be discovered, and it may be superseded by more encompassing theories. But it is still more than a summary of the experimental facts, and its existence transcends its original support. To elevate a collection of facts to a general law invokes taking a leap, and the ultimate being of the law resides in the mind of the physicist.

The legacy

Søren Kierkegaard had expected to depart from life when his *Postscript* was finished. As this did not take place, he had to look around for further tasks to undertake. Not being an impressive figure, with a somewhat ludicrous gait and a weak voice, he had to continue as an influential author. His conviction that one had to take a personal moral stand made him take a look at society and fault it for superficial and insincere conduct. He became the center of strong conflict and violent debate. This stage of his life made his impact on the social scene even stronger, and it lives on as a lasting legacy of his activities. The details of this part of history are not of any essential importance for our arguments here, and hence I will only summarize a few of its features.

One debate with the journal *The Corsair* concerned both ethical and literary matters. Founded in 1840, it was Denmark's first periodical devoted to political satire. It was witty and humorous but in some sense lacked profound sincerity. Its influence on the public was, however, to be strong and it was widely feared in the intellectual circles of Copenhagen. Initially it took a positive attitude to Kierkegaard's activities, but, true to character, he disliked being approved of by such a scandalous medium. There ensued a feud not lacking unpleasant features. The journal ridiculed not only the writings of Kierkegaard but also his appearance, his way of walking and manner of dressing. The writing made Kierkegaard laughing-stock in the streets and left him with a lasting wound of mocked pride. As these controversies started in 1846, they added to his lack of a mission for his activities.

It is curious to note that the role of another satirical journal *Die Fackel* came to catalyze the emergence of Wittgenstein's philosophy in Vienna at the dawn of the twentieth century. This makes one surmise that irony is the most fertile argument in a social debate, its power is felt immediately, but its effects may be huge and lasting.

Kierkegaard's final years were devoted to debate with the Church. He took up this fight as an issue of moral sincerity. As the church of Denmark was a government institution, its clergy were civil servants under the state. For Kierkegaard, only the personally gained relation to the religious could be genuine and true; he found it to be impossible that a church priest could be sincerely religious. This controversial attitude put him in direct conflict with the earthly powers of the Church and came to be one of the lasting memories of his activities.

Kierkegaard, however, had to wait several years before he started this campaign. In January 1854, Bishop Mynster died, and he had been a friend of Kierkegaard's father. It seemed that Kierkegaard wanted to spare him the agony of confronting the son of his friend. After the funeral Kierkegaard lost this restraint and launched his attack. The starting point was the manner in which the Church offered a Christianity devoid of conflict, demands, and sincerity. Kierkegaard respected Grundtvig and his religious movement, but in the light of the new campaign, even Grundtvig was not devoting himself deeply enough to the real implications of a Christian life. Many articles were written, many persons were involved, and much argument was put forward on both sides. Thus the fight was on, and lasted to Kierkegaard's death in 1855.

When dying in the hospital, Kierkegaard had a discussion with his friend Emil Boesen, where he among other things stated: "Greet everyone, I have been fond of them all, and tell them that my life is a great, to others unknown and incomprehensible, suffering." Kierkegaard did not want a pastor at his funeral: "The priests are royal functionaries, royal functionaries have nothing to do with Christianity." This was not to be; Søren's brother Peter Christian arranged the funeral and gave the speech himself. This was on November 18, 1855. The speech was personal, unassuming, and tactful. But it was just what Kierkegaard himself would not have wanted. His view was perhaps better expressed in an unsolicited speech by Henrik Lund, the son of Kierkegaard's sister.

One question still remains to be dealt with. It is often stated that Kierkegaard was mentally ill. He certainly suffered greatly in his mind, and his brother Peter Christian ended his life in 1888 at the age of 83, several years after having succumbed to melancholy and doubts about God at the age of 70. His son Pascal Michael Paul Egede became mentally ill about the age of 30, whereas one of the sons of a sister was very melancholy but died early and escaped any eventual mental incapacity.

We thus see that the Kierkegaard family was of delicate mental constitution, but this may only have been an aspect of their undeniable intelligence combined with an extreme mental sensitivity. Søren himself worked hard, manipulated his public image, and administered his economic inheritance so that it lasted to the end of his life. In spite of all his mental agony and his aversion to rational arguments, there are no signs that he did lose contact with the reality of life. He was eccentric and behaved strangely, but his devotion to thinking and his writing made this mode of life seem almost rational. With another lifestyle, he would have been another person. And as Søren Kierkegaard he lived and acted purposefully, loyal to his task as he saw it.

So what is the legacy Kierkegaard bequeathed on us? His attacks on Hegel's influence were successful, but here he did not stand alone. A generation of thinkers carried philosophy away from Hegel's metaphysical systems and transformed his ideas in various ways. The remaining traces in modern Marxism are tainted with associations which make it hard to discern the Hegelian origin in the theory. The industrialization and social developments have carried society into a realm not fathomed at the time of Hegel. Here Kierkegaard had a very unimportant role to play.

Kierkegaard's attacks on the state church were a lost cause. In many ways, religion is still in the hands of an establishment, more or less tied to the political machinery. The individual is more free to develop his religious persona, but most formalities are still performed by functionaries.

The lasting trace of Kierkegaard's writing is in the literary and philosophical fields. Generations of authors have felt the alienation of a world too strange or too fast for their liking and then the despair and anxiety offered by Kierkegaard has provided an atmosphere where they could breeze freely. They have had to live with his downcast mood, which has driven them to the edge of suicide, but Kierkegaard's influence lingers on in their works.

The twentieth-century European philosophy called existentialism is a direct descendant from Kierkegaard. It has inherited most of his ideas about the position of the individual and his duty to define his relation to the world and God. Whatever the outcome of this personal choice is, the attitude is religious even when it proclaims atheism. It is well documented that the roots of existentialism grew in the soil prepared by Kierkegaard.

Existentialism tends to take a stand against rationalism. This makes it difficult to pinpoint its exact meaning. This, however, is in the spirit of Kierkegaard's work. Ironically enough, his influence has branched over into fields totally outside his sphere of concern. The cultural atmosphere in *fin-de-siècle* Vienna was imbued with existentialist ideas. The impetus to Wittgenstein's *Tractatus* is to be found in this atmosphere, so even if this has been a central text for the development of twentieth-century logistics, it carries with it a part which is pure existentialist discourse. In the later Wittgenstein, this trend becomes a leading one.

Another unexpected influence of Kierkegaard's Danish legacy is the impact on modern physics. In spite of its rationality, this no longer accords with a fully systematic formulation but offers a fractured and hazy world description. Niels Bohr in Copenhagen had grown up in an atmosphere where Kierkegaard's spirit was ever present. It may thus not be an accident that he defined the esoteric framework for the emerging quantum theory of reality. Instead of the isolated individual forming an existential view of his position, modern physics pictures the generic experimentalist as captured in his subjective role of observer, but with a freedom to carry out the next operation at his will. There seems to be little anxiety attached to this position, but the essential amount of uncertainty and limitation corresponds strongly to that expressed by existentialism.

Whether this relation between modern physics and existentialism is necessary or accidental is hard to decide. It may mirror a characteristic of reality or of our mind's way of treating experience, but most likely it suggests a universality in how mind perceives its surroundings. In this case it contains some profound truth about reality and our place in it. Today, however, we lack sufficient insight to formulate this truth. If Kierkegaard was right, it lies in the nature of this truth that we can never know it.

Kierkegaard in modern thinking

The defining feature of Kierkegaard's approach to life is the central role laid by the individual (*den enskilde*). He strongly experienced the plight and duty of the individual person, who is thrown into the midst of his existence and has to find his own way and modes of being and acting. The strong moral core of his thinking is founded on and developed within a frame of Christianity. All aspects of life and motivations for actions (or avoidance thereof) are formulated in the language of religion. At a time when psychology was only an emerging dawn at the horizon, such a framework may have been both natural and efficient. Many aspects of religious discourse have undoubtedly been developed as rules for moral behavior. In fact, lacking direct access to the subconscious machinery of information processing, we may well be able to progress in understanding human mentality using any sufficiently developed area of social intercourse. Freud, on his side, managed to initiate modern psychology by mapping all behavioral patterns on sexual relationships underlying the social structures of society. This view is, from today's perspective, grossly oversimplified and, perhaps, tied to the local environment where it was developed. But, as a method of analysis, Freud's approach has formed a precedent for much later work. We may thus see Kierkegaard's effort as an early attempt to organize the world view of the individual in a situation when religion had come to lack the absolute conviction of earlier times, but still provided a theoretical framework for governing the experiences of living one's own life.

It has been claimed that Kierkegaard's approach is centered entirely on his own personal experience. He understood his subjective concepts as universally applicable, his motivations and choices as characteristic of the human situation. Later existentialist thinkers have started from the same human isolation in a universe lacking certainties and guidelines. However, they have, even in this situation, striven to gain a systematic framework or at least a method to analyze the ever-streaming sensations experienced by the individual. Kierkegaard would have disapproved of any such attempt; he could not approve of propagation of essential personal convictions by rational methods. After the individual has accepted a stand by a leap of "faith", he must live in this conviction to the extent that he is willing to die for it. No rational argument can convey such an absolute certainty. All rational thinkers, philosophers or theologians, err when they provide a formal frame of concepts instead of the absolute convictions implied by the individual's sincerity (*innerligheten*). But Kierkegaard does not claim to be a philosopher, he tends to despise all such academic categories. Yet, looking at the future developments of existentialism, one may wonder if he was not correct; the attempts to form a firm basis for discourse based on the cognitive experience of the individual have not led to any convincing result. Existentialists have not been able to find any Cartesian rock for theory building, based solely on the experience of the individual human stream of consciousness. This is the dilemma motivating Husserl's lifelong struggle with objectivity. Later existentialists seem to have totally abandoned any attempt to understand reality. They have concentrated on the subjective experience and its manifestation in textual material. Relations to scientifically analyzable facts seem to play no role.

In considering Kierkegaard's version of existentialism, I find that his language foreshadows both the linguistic scepticism of Wittgenstein and the observational uncertainty of Bohr. This is best seen in one of Kierkegaard's late works which is presented in some detail below.

The Kierkegaard text most clearly heralding the modern views is the work [28] *Unscientific Postscript* published in 1846. Characteristically, this work purports to have been written by a Johannes Climacus, and Kierkegaard only appears as the publisher of the text. The figure Climacus is based on a real person, a Greek monk living in the Sinai monastery during the sixth century. He wrote a text based on the ladder of James in Genesis. The book, *Scala Paradisi* displays thirty steps to the perfect rest in God. Kierkegaard, no doubt, saw his "stages on life's way" as a latter-day analogy to the method of the original Climacus. He, however, deviated from his master Hegel in abandoning all systematic ways of progress. The step from one rung of the ladder to the next had to be made through a personal commitment, the leap of faith advocated by Kierkegaard.

In *Unscientific Postscript* Kierkegaard explicitly starts from the condition of existence. Basing himself on Lessing, he states that the road to religious conviction emerges from the individual person not from any closed system. No formal structure can encompass the totality of reality. Stating clearly what is posited and assumed, a philosopher may construct a system of thought. No such system may, however, capture the genuine being, which is always open and under eternal development. The reason, according to Kierkegaard, is that the thinker cannot extricate himself from the process of evolution, he may not form a closed objective view from the outside where he cannot go. The objective philosophy cannot reach the inner person who strives to understand the evolution of his cognition. Here Kierkegaard clearly anticipates later developments both in psychology, existentialism, and quantum physics.

In this philosophically oriented text, Kierkegaard states that the goal for the individual is to become subjective. To be objective may be sound within science, but it offers only a cul-de-sac for the understanding of one's personal place in life. He goes as far as to state that only the personal can be true. Only what we know can be designated as known. This use of the term "true" differs, of course, greatly from its use in modern logical parlance. Set in a religious frame, it seems to base its validity only on revealed truth, but in an interesting manner it foreshadows the concept of pragmatic "truth" advocated by William James not too much later.

Kierkegaard clearly contrasts the objective thinking from the subjective [28]:

> The objective thinking does not busy itself with the thinking individual and his existence. The subjectively thinking person, on the other hand, is engaged as a human being existing in his own thinking. This implies a special kind of reflection: the sincerity, the reflection of

inner possession, whereby the thought belongs solely to the thinking person and to nobody else. The objective thinking states all in finished results and lures the whole humanity to take a free ride by offering phrases of answers and conclusions: The subjective thinker rests on being and eliminates the conclusions, partly because he has already accepted this by finding the way and partly because human existence is eternally evolving.

The world view acquired by the subjective is complete and personal. He realizes the impossibility of a common view of total being. He has seen through the web offered by objectivity and even everyday language:

One states something, and the other believes he agrees quite according to the wording. They think they are in agreement and that they understand each other. In case the speaker is unaware of the duality of thoughts, he is also not conscious of the dual meaning of the statement. He cannot imagine that this agreement can be the most profound misunderstanding.

The objective thinking is the method of science, but its relation to the inner person is lacking:

The objective way leads to abstract thinking, to mathematics and historical knowledge of various kinds. It leads ever away from the person whose being or nonbeing becomes something totally irrelevant, which objectively speaking is fully in order.

But also: "The individual, as existing, is unable to follow both ways simultaneously." Kierkegaard here announces a kind of complementarity in psychology. He is, however, well familiar with the objective claim for truth, but he prefers to pursue his own version as truth being based on the subjectively collective experience of the individual:

Empirically ie according to experience, one determines the truth as agreement with reality. Already thereby, truth becomes something eternally becoming, something one desires and aims at, because the goal of one's experience is never finished, and because the existing having the experience is himself eternally becoming. Thus truth consists of an endless approaching and correcting, whose beginning cannot be established because there cannot be any established end. Its beginning cannot be a hypothesis either, because a hypothesis always implicates that a conclusive decision is conceivable.

Heuristically approached, this may well be taken to describe the continuous process of extracting knowledge about empirical reality including the procedures of quantum physics. Kierkegaard assigns truth only to the subjective conviction achieved by faith:

Objectivity holds only the fixation of thought, subjectively the inner significance, the sincerity. In its most intense form, this *how* is the manifestation of devotion, which emerges in a perspective *sub specie aeternitatis*, which is the truth. But this devotion is just the subjectivity. Thus subjectivity is the truth. . . .

In a mathematical expression, on the other hand, objectivity is given, but then its truth is only an uninteresting truth.

Does this foreshadow Wittgenstein's view that mathematical validity is only inconsequential tautology? Individuality, on the other hand, is not easily discerned in a fellow communicator:

When a thought acquires its proper verbal expression, which occurs through immediate reflection, there appears promptly a second reflection: how is this statement related to the existing human person who utters it, and how does it mirror his own relation to the thought itself?

When applied to one's own utterances and the corresponding thought, this attitude reflects the enigma encountered by Møller's Danish student.

3

THE VIEW FROM COPENHAGEN

3.1. Bohr the master

3.1.1. Life and position

There is no doubt, that Niels Bohr was the central figure forging the world view of the emerging quantum description of empirical reality. Not only did he initiate the revolution in atomic physics, but he also formed the center around which the tribe of younger physicists gathered in brainstorming sessions whose ideas were subsequently worked out separately or in groups. All this was orchestrated by Bohr, who also acted as a godfather and sounding board for the actual hard task to produce the technical details. His sphere of influence has been referred to by the fuzzy epithet "Copenhagen interpretation". I will later go into some details of this school of physics, but first we need to establish the family and culture which made Niels Bohr the singular individual he was.

In 1881 Christian Bohr married Ellen Adler. Christian descended from a German family that had moved to Denmark. The family had a tradition of learning and teaching, which was manifest by the life of Christian. He studied medicine, did research on the physics of gases, and developed a theory for respiration and oxygen utilization. Ellen descended from a family of Jewish origin which had pursued business in Hamburg and London. Her parents moved to Copenhagen in 1850, and the father eventually developed into one of the wealthiest men in Denmark. The Jewish origins remained a decisive influence on the life of the family. Ellen had three brothers and two sisters.

Niels Bohr was born on October 7, 1885 as the middle child in the family. The elder sister Jenny was born in 1883 and the younger brother Harald in 1887. He was to become a successful mathematician and an international level football player.

In 1890 Christian Bohr was promoted to the status of full professor in physiology at the University of Copenhagen.

All three Bohr children were babtized into the Protestant faith.

In 1903 Niels entered the University of Copenhagen. Christian Bohr was Rector 1905–6 at the university.

In 1909 Niels met Margarethe Nørlund, his wife-to-be. Margarethe was the daughter of a pharmacist from Slagelse, and the couple met because her two brothers were friends of Niels Bohr's.

Niels suffered the loss of his father who died at the young age of 56 in 1911. Niels also acquired his Ph.D. in this year. He left Denmark to work with

J. J. Thomson in Cambridge, England. The collaboration did not work out, and Rutherford arranged to have Bohr transferred to Manchester.

In 1912 Niels Bohr worked in Manchester, left the Lutheran Church, and returned to Denmark, where he married Margarethe.

He published his theory of the hydrogen atom and gave lectures on this theory in 1913. He also showed that β- decay is a nuclear process.

In 1914 he returned to Manchester as lecturer but received a professorship in Copenhagen and returned to Denmark.

The periodic system was explained by Bohr in 1920–2.

The theoretical physics institute in Copenhagen was opened in 1921. This was to become the famous Niels Bohr Institute.

In 1922 Bohr lectured in Göttingen at what has been called the "Bohr Festspiele". Here he met Pauli and Heisenberg. Bohr received the Nobel Prize for "Investigations of the structure of atoms and the emission of radiation emanating from them".

First papers on quantum physics by Heisenberg and Bohr appeared in 1925, and the theory was developed by Heisenberg and Jordan. Uhlenbeck and Goudsmit introduced the electron spin as a fundamental entity.

In 1926 Schrödinger published his first paper on wave mechanics. Born introduced the idea that quantum mechanics predicts probabilities. Bohr discusseed the foundations of quantum theory with Heisenberg and Schrödinger in Copenhagen. This was the famous occasion when the strain and intensity of the arguments forced Schrödinger to bed from exhaustion. Also many of the other key figures of quantum physics, for example Kramers (the Netherlands), Klein (Sweden), and Dirac (England), visited Bohr that year.

In 1927 Heisenberg published his paper on the uncertainty principle associated with his name. This was written in Copenhagen. Bohr presented the concept of complementarity at the Volta meeting in Como. In Brussels the Fifth Solvay conference took place. Here the famous series Bohr-Einstein debates started.

In the year 1930 physicists met at the Sixth Solvey conference and continued the Bohr–Einstein dialog. On this occasion the gravitational effects of the clock-in-the-box were considered. In 1934, Bohr visited the Soviet Union for the first time. He lectured in Moscow, Leningrad, and Charkov.

Bohr's eldest son Christian drowned in 1934. The Bohrs had six sons, of which Aage born in 1922 became a successful physicist and winner of the Nobel Prize.

The publication of the Einstein-Podolsky-Rosen paper (=EPR) and Bohr's rebuttal took place in 1935. In 1938 Bohr defined his concept of a "phenomenon" and related complementarity to human culture.

After the German occupation of Denmark, April 9, 1940, Bohr's situation was precarious. This was the year of the controversial visit of Heisenberg to Bohr's institute. The actual discussions and their motivations have been debated endlessly even up to the scenes of theaters.

In 1943 Bohr and his wife fled Denmark to arrive in the US via Stockholm and London. He was only to return when Denmark was liberated.

1962: Niels Bohr died of heart failure on November 18.

These dates indicate the central milestone's in Bohr's life and the physics events where his influence was most decisive. In addition to these instances, he contributed greatly as an administrator and supporter of experimental physics. He also became an influential spokesman for peace and international harmony. These activities are left out above because they have less bearing on the arguments pursued here. For further details references should be consulted.

3.1.2. Background and influences

Bohr never devoted any time to systematic study of philosophy. In his writings he does not refer to any of the great authorities. Still, much of his contribution to the development of modern physics was to strive to clarify the possibilities and limitations of our means to describe observations, a typical problem of philosophy usually referred to as *epistemology*. However, when late in life he was asked about his interest in the discipline he is reported to have answered: "It was, in a way, my life." He, however complained that no philosopher ever understood his concept of *complementarity*.

Early Bohr seems to have contemplated the possibilities of attaining meaning and the ways to avoid ambiguity. Already as a boy, he enjoyed the privilege of listening to advanced discussions about the meaning and significance of science. His father, the professor of physiology, collected his peers at home for discussions. Among these were Christian Christiansen, the professor of physics at the university, Harald Høffding, the professor of philosophy, and Vilhelm Thomsen, a famous linguist. The Bohr boys sat in on these discussions, and they were, no doubt, imbued with much of the cultural debate which flourished in Copenhagen in those days. Thus they became educated without attending lectures, instructed in dialectics without study, and used to logical argumentation without formal training. This situation, however, makes it difficult to ascertain when Niels learned about various ideas in philosophy.

Two of the participants later served as Niels Bohr's teachers. Christiansen, as the single professor in physics at the university, taught him both experimental and theoretical physics, apparently both efficiently and successfully. As an undergraduate student, Bohr had to take a compulsory course introducing philosophy. Here he, no doubt, got acquainted with the classic names and ideas of the academic tradition in philosophy. He also learned to respect and admire the teacher, Høffding, for his clarity and openness of mind. The respect must have been mutual, because Høffding let Bohr read a manuscript of his textbook on logic while Bohr was still an undergraduate student. His possible role as Bohr's source of philosophical knowledge justifies a scrutiny of his intellectual position, which I will undertake below. Later Høffding was trying to understand Bohr's concept of complementarity, but without great success.

Niels Bohr's younger brother Harold became a mathematician. He, however, also was a member of the 1908 Danish Olympic soccer team. When he took the doctoral degree, this was announced on the sports page. Thus Niels early got to know the details of advanced mathematics.

Harold Bohr came to Göttingen in 1909 to study with Hilbert. When Niels visited him, he had the opportunity to meet Hilbert and discuss modern mathematics as represented by the Göttingen school. Niels visited his brother frequently and informed Hilbert about the progress in contemporary physics. Thus Niels, no doubt, got some connections with the modern developments of mathematics, which must have helped him in grasping the more formal aspects of quantum physics.

Niels Bohr seems to have had an uncanny ability to attach philosophical significance to formal machinery. This is a trend clearly exhibited by his discourse as a mature scientist, but already as a student, he was attracted by such relations. In mathematics, he learned about the way a complex function may acquire different values on different sheets of its definition. These are the well-known Riemann sheets, which allow a function to present different results at seemingly identical points in the plane. The function itself is defined on the collection of all such sheets. Bohr saw in this an analogy with language: The use of words may carry different meaning depending on the setting of the discourse. The similarity with his later ideas about complementarity seems obvious.

Another philosopher who appears to have been important for Bohr was the psychologist William James (1842–1910). It is not clear to what extent Bohr had read the works of James nor when he may have done so. However, James consistently stresses that the description of mental concepts cannot remain unaffected by the methods of investigating them. The wording is close to that which reappears in quantum theory later. Also his concept of flow of consciousness versus external experience seemed to support Bohr's combination of phenomena from observational setting and microscopic atomic processes. Bohr introduced the novel concept of a *phenomenon*, as one consisting of both the observed object, the unambiguous result of the observation, and the full description of the conditions validating the result. This new concept was to form the foundation for the utilization of the approach of *complementarity*, and it is one of Bohr's essential contributions to philosophy. Later he was to apply this concept also outside its original setting as a description of quantum measurements.

Bohr was clearly aware of the Viennese movement of *logical positivism*. He had met Otto Neurath already in 1934. Furthermore, in 1936 the "2. International Kongress für Einheit der Wissenschaft" took place in Copenhagen. This was organized by Otto Neurat and Jørgen Jørgensen, the latter being a Danish representative of the Viennese movement. In addition to these two, Bohr had contacts with Philipp Frank. Bohr exchanged letters with Neurath over the following couple of years, but it is difficult to judge how strongly the positivist aversion against metaphysics influenced Bohr's view of reality. He remained a realist concerning the objects of study in physics, but an instrumentalist concerning their observable attributes.

Bohr must at some time have devoted reading to getting acquainted with oriental philosophy:

We must in fact turn to quite other branches of science, such as psychology, or even to that kind of epistemological problems with which such thinkers like Buddha and Lao-Tse have been confronted.

Significantly, when, in 1947, Bohr chose his coat of arms, he adopted the Chinese Yin-Yang symbol and the device *contraria sunt complementa.* Do we here trace the origin of the concept *complementarity* to the logical relation of "contraries" which are not "contradictory"?

Nobody can fathom exactly where Bohr acquired his information about philosophy. He read, but not extensively; he wrote only slowly and with difficulty. His manuscripts were rewritten and reworked numerous times, and the final product was often so contorted that its message is hard to extract even today when the quantum theory is firmly established. Bohr also wanted to work mainly orally. He needed young knowledgeable disciples to explain the issues, and he used to produce his texts by dictating to an assistant. When he understood an issue, his grasp was full and deep, but the formal expression of his vision was often beyond the capacity of the verbal tools at our disposal.

3.1.3. References relating to Bohr

There is a multitude of writings dedicated to the ideas and opinions of Niels Bohr. Here I only want to direct the reader to those sources on which my present summary is based. They cover the more philosophically oriented aspects of his thinking; for the physical analyses other sources must be consulted.

A summary of Bohr's argument is given by Honner [29], where also a historical review is offered. Honner stresses the transcendental character of Bohr's concept of empirical activity of physics. Murdoch [30] analyzes the same material and concludes that Bohr's attitude is strongly similar to that of "pragmatism". Folse [31] summarizes the received view of the s.c. Copenhagen school. Pais [32] inserts Bohr's achievements into a wider cultural perspective. Faye [33] goes through Bohr's background and its relation to philosophy. Here the details of Høffding's philosophy are presented. A more recent text by Plotnitsky [34] tends to stress the anti-realist interpretation of Bohr's work.

Bohr's own writings on philosophy and epistemology are collected in four booklets [35] to [38]. The impact and essential importance of Niels Bohr can be obtained from the published conference [39].

3.1.4. The life and opinions of Høffding

Biographical résumé

Harald Høffding was born on March 11, 1843 as the son of a well-to-do merchant. The family home was situated in the very center of Copenhagen. The home of the Kierkegaard family was not far away.

Høffding was born in the same year as the famous Copenhagen Tivoli was opened. This was originally situated outside the city gates, but the growth of the city and its various activities made Tivoli a central part of the emerging modern city. When Høffding died in 1931, Copenhagen had essentially achieved its present status.

Høffding entered university in 1861 and started his studies in theology and philosophy. Following the tradition that talented students entered the clergy, he aimed at the theological degree which he took in 1865. However, his logical and psychological studies under F. C. Sibbern had opened his eyes to philosophy. His favorite authors here became Spencer and Comte. An even more influential factor in his decision to devote himself to philosophy was the contact with the writings of Kierkegaard. He attended lectures on Kierkegaard's writings, but also read his works.

The critical attitude towards religion displayed by Kierkegaard made it impossible for Høffding to seek a living as a priest. Especially the text *Concluding Unscientific Postscript* awoke Høffding's interest to the relation of personally centered subjectivity and truth.

During the years 1868 to 1869, Høffding lived in Paris and studied French and English philosophy. Here he got his firm ground in positivism and empiricism even if he never became merely a disciple. He formed a strong personal view of philosophy, which led him to contribute to a multitude of questions during his long life. He, however, never took much interest in formal logic, where his understanding was not at the highest level.

In 1870 Høffding married Emma Pape, who, however, died in 1877. At the age of 81, he married Greta Ellstam,—Fifty-seven, year his junior who died in 1930, one year before Høffding himself.

Høffding became a university professor in 1883 and retired in 1915. He was president of the university 1875–82. During the First World War, he worked devotedly for the Danish Red Cross, and even in retirement he remained active and produced written documents up to his death. He treated many philosophical questions related both to theological and moral issues. His last interest was in epistemology and the theory of knowledge, which was strongly related to his contacts with Niels Bohr. As a curiosity, it may be mentioned that, upon retirement, Høffding became the first honorary inhabitant in the apartment provided by the brewery Carlsberg to house eminent Danish scientists. After Høffding's stay, Niels Bohr became his successor in the very same apartment.

Høffding's intellectual influence on youth was significant. He was here competing with the widely known radical *littérateur* Georg Brandes. Brandes's position at the university had been a drawn-out struggle between his supporters and representatives of the conservative establishment. Denmark was in dire need of having a Nobel Laureate in literature, but Brandes was too controversial. Thus Høffding was nominated a few times in 1911–16. The powerful Carl David af Wirsén represented a highly conservative attitude in the Swedish Academy, and the "modern" approach by Høffding could not be accepted. The situation was not improved by Høffding's reputation as a writer of formalistic and difficult prose.

Høffding's philosophy

The cultural atmosphere in Copenhagen was strongly dominated by N. F. S. Grundtvig (1783–1872). He accepted the contemporary German criticism of the origin of religious canon. He advocated a personal approach to faith, but regarded the Lutheran challenge to the individual to be too much influenced by papal thinking. By stressing the importance of personal religious experience, he managed to convey a strong religious feeling to the common people of Denmark.

Combining religious thinking with a powerful emphasis on national identity, he managed to unify two trends in the contemporary intellectual debate. He singlehandedly seems to have created the concept of the Danish people. His influence on his country represents a unique case history in Europe.

At the beginning of the nineteenth century, the philosophical tradition in Copenhagen was thus strongly dominated by religious issues. This tendency lingers on in the development of the debate created by Kierkegaard's writings. The resulting development in philosophy then led to an increasing emphasis on psychology. This is not surprising, because psychology emerged as a separate academic discipline of study only towards the end of the century. However, in Denmark this orientation became a bulwark against the strong influence from Hegel. Its concentration on the individual also prevented the Romantic philosophical tradition from gaining strong support.

The dominating influence in Copenhagen derived from F. C. Sibbern (1785–1872). Becoming professor of philosophy in 1813, he remained at the position for fifty-seven years. Having started under the influence of the Romantic philosophy represented by Schelling, he developed into an early forerunner of the existentialist philosophy. His influence on Kierkegaard and Høffding was decisive. Both rejected the adoption of great systems and founded the view of the world on personal experience. Sibbern seems to have created the term "the individual" ("den enskilde") as later to be used by Kierkegaard. In the debate about "belief and knowledge", Kierkegaard came to stress the paradoxical in faith which necessitates the "leap of faith".

Sibbern's philosophical colleague at the university was Poul Møller (1794–1838), who wrote little on philosophy, but whose work on the Danish student exerted a decisive influence on Niels Bohr. The existentially and subjectively oriented philosophy of personality advocated by Møller and Sibbern opposed the influence of Hegel and formed the platform for the extensive and influential work by Kierkegaard. In contrast to the absolute picture presented by Hegel, the philosophy of the individual offers a view of continuous development harmonious with the progress of the individual in the world. There was an attempt to reconcile Hegel's approach with religion by J. L. Heiberg and H. L. Martensen. These attempts were strongly opposed by Kierkegaard, who regarded their dogmatism as incompatible with the subjectivity dominating the philosophy in Denmark.

Høffding and Bohr

After his initial interest in moral and faith, Høffding turned to more scientific issues later. In 1875 he started his study of psychology, which resulted in a book in 1882. Here he related the theoretical discourse to experiments and foreshadowed the emergence of *Gestaltpsychology*. Following his conviction of personal subjectivity, Høffding does not consider knowledge as accumulated experience but insight into the interrelationships between experiences.

Høffding kept in contact with William James, who shared many of his opinions. In 1902 he published a book on *Problems of Philosophy*, which was issued in English in 1905 including a preface by James. The book Høffding published in 1904 on modern philosophy includes the ideas of Mach, Hertz, and Ostwald, showing his interest in modern physical science.

Høffding's interest in epistemology is documented in the work *Human Thought: Its Form and Its Tasks* published in 1910. These issues were the topics of informal meetings organized in the home of Niels Bohr's father. Here the young Niels acquired many of the philosophical attitudes that came to shape the formulation of quantum physics. The interdisciplinary character of the meetings was guaranteed by the combination of the participants:

- Christian Bohr (physiologist)
- Harald Høffding (philosopher)
- Wilhelm Thomsen (linguist)
- Christian Christiansen (physicist and Niels's teacher).

The discussion group used to meet after the Academy meetings up to the death of Bohr senior in 1911. Following his father's example, Niels entered in 1905 a discussion group EKLIPTIKA after starting at the University in 1903. This group, of slightly varying composition, discussed Høffding's lectures, and hence we may conclude that Niels had attended them by then. Some members of the group may have met as late as in 1916.

Edgar Rubin was the driving force of the EKLIPTIKA. He later founded a "Copenhagen school" of psychology. This went beyond the Gestalt approach looking at the mind as an integrated totality. He presumably drew Niels's attention to the work by William James. The interest of the group thus seems to be strongly under the influence of Høffding.

It is slightly unclear when Niels Bohr read James. In old age, he claims to have read about "the stream on consciousness" already as a student. Rosenfeld suggests that this happened only in 1932. These statements are not necessarily contradictory; Bohr may have learned about James as a student but read him more carefully only later. He also benefitted from Høffding's lectures on Kierkegaard. It seems clear, that Bohr read only little original philosophy, but he got acquainted with its main ideas from lectures and discussions.

Kierkegaard seems to have had an essential influence on Bohr; this was hard to avoid in Denmark at that time. According to Kierkegaard, we may create a system of ideas but not of reality. The personal approach is here confronted with the Hegelian approach. The concept of a "leap" utilized by Kierkegaard implies that the state changes in a moment; what happens in the transition cannot be described. According to Høffding such discontinuities play a role in the scientific description of reality. He explicitly refers to radioactive decay and mutations. According to him this introduces an element of "irrationality" into the description. These features may have been discussed with Bohr as a part of the "subject—object" discourse carried out by Høffding.

When Bohr later presented his concept of "complementarity", first in the Como lectures, his choice of language seems strongly influenced by Høffding. After Bohr had explained his complementary view of physics, he was told that "you said the same 20 years ago" by a collaborator (this story is referred either to Kalcker or to Heisenberg?).

Høffding's last work was an essay on the theory of knowledge. He stresses that every concept expresses a relation. In psychology this introduces a complementarity between causal implications and the feeling of freedom of the will. According to him, our ability to understand is essential to human thought. Cognition depends not only on the phenomena but also on our entire intellectual apparatus of understanding. In contrast to Kant, he claims that our concepts are not *a priori* but develop. They reflect our capacity for organization of growing experience.

Høffding does not present a formal definition of "truth", but he characterizes this as "dynamical" because it emerges from the activity of the intellect. Here his relationship to "pragmatism" is obvious. Truth cannot be a correspondence with reality because we only know reality through our cognition. Not every element of experience can be made to conform to our comprehension of reality. The truth of a principle rests in its utility. It is useful if it brings together as many phenomena as possible. "We have only knowledge about a thing as far as it is a cause or an effect; every property is a relation".

The concepts of theories are ideals, not found in the world of experiences. Causality is such a concept. Not every element of reality can be made to conform to the categories of our understanding. We know the world through ourselves and ourselves through the world. Romantics develop the former and neglect the latter, the positivists stress the latter and disregard the former. No final goal for our knowledge can exist. The world can only be understood in its "wholeness"; this is a fundamental category for Høffding. It cannot, however, explain its elements; neither is the opposite possible. Høffding here anticipates the later concept of "emergence".

Attention in itself changes the state to which it is directed, to which it is intended. Thus the observer and the observed cannot be kept separate. Høffding arrives at this conclusion from psychology, where the analyst and his patient affect each other mutually, an idea only later utilized by Freud.

Høffding introduces an "identity hypothesis" between body and mind. They do not coincide, neither do they cause the states of each other. They are not acting in parallel either, rather their ontological status is the same; this is his "double aspect theory". They form "two languages which can be translated into each other, but they cannot be derived from each other". For understanding we need both. Physiology and psychology are both necessary but independent descriptions. They are complementary. Materialism and vitalism are equally useless for explanations.

Høffding introduces the concept of "analogy" as "similarity of relations". Here he anticipates the modern concept of a model for some set of experiences. This idea he seems to have learned from reading Maxwell. According to Høffding, the application of numbers to items is based on an analogy between series of numbers and series of items. This conclusion should remind us of the Gödel approach to proof formulations. In natural sciences, we apply mathematical structures to webs of phenomena. In the early publications Niels Bohr calls his theory "a formal analogy" with old electrodynamics:

> Under these circumstances one must naturally constantly bear in mind that one is operating with analogies, and this step, in which the applications of these analogies is limited in each case, is of decisive significance.

Is this a version of Wittgenstein's picture theory of descriptions?

3.2. Bohr's legacy

3.2.1. Bohr's conceptual analysis

In this section, I will summarize some of Bohr's ideas that seem most closely related to Høffding. It is a strenuous enterprise to disentangle the various turns of his thinking. Honner [29] expresses this eloquently: "May I also warn that Bohr's thought is notoriously elliptic, and the survey of different aspects of his vision demands what may seem to be an inordinate number of treks across the same ground." This should be kept in mind when the present text is read; it is my personal reading and no final interpretation.

As noted above, there are many features of Bohr's view of knowledge which he inherited from Høffding. Especially the two key concepts "correspondence" and "complementarity" derive to a large extent from him.

The correspondence is a rational generalization of the ordinary theory of radiation. Bohr does not regard it as a "law of nature" but as a methodological strategy. As such it contains no necessary ontological implications. Its only point of contact with human experience is that of classical physics. This is necessary in order to have a uniquely communicable point of reference. It gives us the only reality we can comprehend.

The irrational element enters the theory in the unavoidable discontinuity of physical processes. Here one may wonder if Bohr should not have written "discreteness" instead of discontinuity. He derives this from the experimentally established quantum of action; modern theory provides a more profoundly complicated picture of quantum phenomena.

Much of the enigma centers on the role of the observer, the quantum measurement problem. Where should we place the cut between subject and object, and how is the process of acquiring information affecting the object? The ultimate manifestation of this is found in Møller's tale of the Danish student. Self-observation may prevent the subject from action. Høffding derives these problems from psychology, where any interaction between individuals will leave neither one unaffected. Bohr elevates this question to a universal principle of epistemology. As Bohr understands this, conscious analysis of a physical concept stands in a relation of exclusion to its immediate application. This forces on us the view denoted by complementarity.

The spirit of Kant still lingers over the approach by Bohr. This derives from the contemporary philosophical atmosphere through Høffding. Even if the absolute validity of the categories has been abandoned, the role of human concepts remains much the same. In Bohr they take the name "forms of perception", which governs much of his discourse. This derives from Høffding who also uses the term "forms of understanding".

In the later writings, Bohr utilizes his central concept of complementarity. This denotes two classically incompatible, seemingly contradictory, ways of approach to a concept, where both are necessary for a full description of the phenomenon.

> The phenomenon is an object existing only as an immediate integral experience, it cannot be regarded as an independent object even though the phenomenon, as an observed object, is presented to, not created by consciousness. ([33], p. 167)

This fullness is introduced as a category necessary for the acquisition of knowledge. In psychology, Høffding introduced the "double aspect theory" between physiological descriptions and use of mental terminology. This is not a theory of parallelism but an identification of the two modes of description as intending the same existing object.

According to Bohr, we cannot know the true nature of reality. Applying the physical concepts to objects without interactions, they become meaningless. Our theories do not aim at real essences of the objects but only to determine, as far as the limitations allow it, the relations between the multifarious aspects of our experiences.

The focus then comes on the tools of such descriptions, the language of theoretical systems. In this manner, Bohr changes his view of complementarity from an epistemological approach to a more semantical one. This change may well have been induced by the arguments encountered in the EPR debate. As all we can access is our description, we should be guarded in how we phrase this. The discussion in EPR uses the concepts of "locality" and "separability", but without denoting in which situations these are to be used, they are not unambiguous. Bohr demands that we define the "very conditions giving the predictions". Today we may, perhaps, rephrase this by stating that quantum theory can only give conditional probabilities. Distributions of observed experimental outcomes are conditioned on well-defined state preparations. The phenomenon is not only the object and the observation but the interaction itself serves to determine how the phenomenon is to be described.

Much of the analysis of Bohr's presentation of his views and, in particular, the presentation of the concept of complementarity, suggests that he is a pure positivist claiming at best an agnostic approach to reality. However, this must be an incorrect interpretation of his views; like all working physicists he is ultimately a realist. He constantly refers to the "atomic object" which is described by the theory. We may not know the essence of this entity, but our empirical information tells us how to achieve a valid description of its response to our probing. When all available information is accessible, we may constitute a phenomenological object, which ontologically corresponds to "features of reality". This information and its modes of acquisition constitutes in some manner real knowledge about our ambient nature. But such epistemological method can only be based on a careful and consistent use of semantics.

3.2.2. Bohr's view of physics

The view from 1929

Observations

In [35] Bohr summarizes his understanding just after the emergence of the formal apparatus of quantum theory. Here I briefly review his arguments.

The task of science is to extend the range of experience and reduce it to order. The basis is the language developed in our natural surroundings. We coordinate the experience with the external world. Then the existence of the quantum of action imposes on the individual process an irrational element of discontinuity quite foreign to classical physics. Bohr always started from the fact that action cannot change continuously. This conviction may derive from his 1913 solution to the atomic stability problem.

Essential is the combination of phenomena with their observation. The latter must be described using "forms of perception" which are idealizations. The forms of perception offer the general conditions underlying the creation of man's concepts. We may here see how Bohr is influenced by Kant's categories. These were presumed fixed and unchangeable. Atomic physics has taught us how, without leaving common language, it is possible to create a framework sufficiently wide for an exhaustive description of new experience. Bohr repeatedly takes the example of the relativity revolution, which forced us to re-evaluate our inherited views of space and time.

Consequently, we cannot dispense with those forms of perception, which color our language and in which our experience must ultimately be expressed. Quantum theory, however, defies strict causality but introduces the new mode of "complementarity". This implies that the use of some classical concepts precludes the use of other classical concepts, which in a different context are equally necessary for the full elucidation of phenomena.

Observation introduces an asymmetry between past and future, which prevents sharp distinction between the phenomenon and the agency of observation. Hence it is not surprising that only statistical regularities between causes and observations prevail. Here Bohr seems to imply the necessity of a statistical description only on the basis of the impossibility of a classical description in the quantum domain. This conclusion does not seem to be logically unavoidable.

Reality

Bohr does regard the objects studied in physics as "real". He states: "Every doubt regarding the reality of atoms has been removed, and we have knowledge even of the inner structure of atoms." We should remember that, when this was written, the school of Mach had quite recently denied the reality of atoms.

As to the reality of the concepts used, Bohr states: "The stationary states possess as much or as little 'reality' as the elementary particles themselves." Thus both the quantum objects and their property to be in states are part of the same "reality". All physical attributes do not, however, allow an objectification in a classical way.

The theory provides no guide to understanding the existence of elementary particles with their mass and charge. Thus the theory offers epistemological recipes only, and carries no (necessary) ontological implications. The situation is new: The irrationality of such a description cannot be eliminated by replacing classical concepts by new conceptual forms. The classical concepts used in describing the phenomena may retain their "meaning" even when they do not take numerical values. As an example I would mention the "conservation laws", which are found to hold even in individual quantum processes. They, however, preclude the simultaneous description of these processes in space and time.

The use of classical concepts in quantum theory derives from their relations to abstract symbols. The theory rests on a formalism which leaves room for the quantum of action. The mathematical apparatus should here not be considered a separate branch of knowledge but it may be seen as a refinement of general language. As an empiricist, Bohr could not be happy with a notion of physics which was purely theoretical; he could never accept any possibility of constructing a new language for quantum physics.

The introduction of mathematical symbolism sets up an ideal of objectivity with scarcely any limits as long as we remain within a self-contained field of applied logic. In natural science, however, we have no such strictly self-contained field of applications because the formalism must adapt to new facts requiring the revision of the fundamental concepts. They may be used unambiguously because of the consistency of the classical theory. However, the uncertainty relations set limits to the "meaning" of the information carried by quantum attributes. The purpose is not to disclose "essences" but to track down relations in experience. Words like "to be" and to "know" lose their unambiguous meaning.

In this light, I may here suggest that the Heisenberg microscope is a consistency check: Assuming that classical theory applies, even then the perturbation by observation prevents the violation of the quantum uncertainties. This guarantees the consistency of the classical limit.

Associative thinking and presentation of personality exhibit, according to Bohr, a suggestive analogy with the wave description and the motion of elementary particles governed by superposition and their indestructible individuality. Do we hear echoes of James's presentation of consciousness here?

The nature of our consciousness brings about a complementary relationship in all domains between the analysis of a concept and its immediate application. Bohr states: "We have encountered a problem capable of reminding us anew of the general conditions underlying all human understanding, which from time immemorial have attracted the attention of philosophers."

The Como lecture 1926–1927

This is the famous lecture where Bohr introduces the concept of "complementarity". The argument much follows what was presented above. I quote some of the key excerpts ([35], p. 52):

The steady state of a system prevails only in the absence of all interactions, but then we can extract no information about it. To measure the state, we have to accept certain interactions, and then a unique description is no longer possible. The quantum postulate of discreteness forces one to choose which classical description is to be used. The limitations of the classical description are expressed in complementary ways. Complementarity is not just a relationship between modes of description but also a mode of description itself. The procedure captures the difficulty in the symbolic method. This can be regarded as a rational generalization of classical theory. We can see this by applying Bohr's "correspondence principle". This does not overcome the quantum postulate but derives from it.

Later developments

Physics and language

In the later writings [36] Bohr applies his epistemological view of knowledge to ever more general situations. He discusses the character of life and the processes of society. Here I want to present only his view of man's place in his social context, because this throws light on Bohr's world view as emerging from his intellectual background and experience from quantum physics.

According to Bohr, the epistemological lesson from science makes it impossible to distinguish sharply between natural philosophy and human culture. Far from containing any mysticism foreign to the spirit of science, the notion of complementarity points to the logical conditions for description and comprehension of experience in atomic physics. The aim of every physical experiment is to obtain knowledge under reproducible and communicable conditions. "This implies the use of everyday concepts refined by the terminology of classical physics."

No proper human thinking is imaginable without the use of concepts framed in some language which every generation has to learn anew. Use of language stands in an exclusive relation of complementarity to the display of inherited instinct. As Bohr says: "We are suspended over a bottomless pit caught in our own words." Every child is born with a predisposition to fit into the culture of its environment. Formal logic is no longer sharply distinguished from semantics and even philosophical syntax.

A special role is played by mathematics. It is introduced as a refinement of general language representing relationships for which ordinary language is imprecise or cumbersome. It eliminates the reference to the "conscious person", which infiltrates daily language.

All knowledge presents itself within a conceptual framework adapted to account for previous experience, and any such framework may prove too narrow to comprehend new experience. In contrast to Kant's *a priori* concepts, Bohr claims that his are derived from observed facts. Study of a wide field of physical experience has revealed the necessity of a closer consideration of the observational problem. The impossibility of providing an unambigous content to the idea of subconsciousness corresponds to the impossibility of pictorial interpretation in quantum mechanics. Every conscious experience leaves an irreversible recording in the nervous system not open to introspection. We cannot know why we are what we are. Not even the communication between individuals separates animals from men, albeit our speech places us in an essentially different situation.

The essential wholeness of a proper quantum phenomenon finds logical expression in the circumstance that any attempt at its well-defined subdivision would require a change in the experimental arrangements incompatible with the appearance of the phenomenon itself.

Observations and reality

The material basis for physics is real. Bohr never argued that the notion of an independent reality is meaningless. The success of quantum theory removes the last trace of the old belief that the coarseness of our senses would for ever prevent us from obtaining direct information on individual atoms. Since Bohr's days, much information of this kind has been gathered.

A scientist cannot but assume that there is a reality existing behind experienced phenomena, but he must take a lot of care in what he says about it. It is essential to understand that an experiment concerning a phenomenon which in principle lies outside the range of classical physics can be interpreted as giving information about independent properties of the object, but the choice of instrument also enters as an unavoidable component. The use of language is essential not only in manipulating and constructing the measuring instruments but also in describing the actual experimental result. The separation between the object and the measuring device can be chosen arbitrarily, but only within the region where the description of the process is effectively equivalent with a classical description.

Phrases like "disturbance of phenomenon" and "creation of physical attributes" are apt to cause confusion since the words "phenomena" and "observation" are used in a way which is incompatible with common language and practical definitions.

Use of the word "phenomenon" is to be restricted to descriptions including an account of the whole experimental arrangement. An atomic phenomenon is "closed", when based on registration by means of suitable "amplification" devices with "irreversible" functioning. The information we have obtained in a measurement cannot be unlearned. Complementarity does not involve a departure from our position as detached observers. Because different individual processes may take place in a given experimental arrangement, the results are of an inherently "statistical character". Concerning the relation of object to the apparatus reading the observation, Bohr writes ([38], p. 82):

> It is true that the place within each measurement procedure where this discrimination is made is in both cases largely a matter of convenience....
>
> Its fundamental importance in quantum theory, as we have seen, has its root in the indispensable use of classical concepts in the interpretation of all proper measurements, even though the classical theories do not suffice in accounting for the new types of regularities with which we are concerned in atomic physics....
>
> By securing its proper correspondence with the classical theory, these theorems exclude in particular any imaginable inconsistency in the quantum-mechanical description, connected with a change of the place where the discrimination is made between object and measuring agencies. In fact it is an obvious consequence of the above argumentation that in each experimental arrangement and measurement procedure we have only a free choice of this place within a region where the quantum-mechanical description of the process concerned is effectively equivalent with the classical description.

This provides a brief version of Bohr's final view of "measurements". It defines our position with respect to analysis and synthesis of experience. Concepts like "realism" and "idealism" find no place in an objective description. Even more explicitly ([38], p. 100):

> We must recognize that a measurement can mean nothing else than unambiguous comparison of some property of the object under investigation with a corresponding property of another system, serving as a measuring instrument, and for which this property is directly determinable according to its definition in everyday language or in the terminology of classical physics. While within the scope of classical physics such a comparison can be obtained without interfering essentially with the behaviour of the object, this is not so in the field of quantum theory, where the interaction between the object and the measuring instrument will have an essential influence on the phenomenon itself.

Bohr here recognizes the need for "pointer variables", i.e. quantum systems whose state can be monitored unambiguously.

Discussions with Einstein

Bohr developed many of the basic quantum concepts in disputes with Einstein. Einstein presented *gedanken experiments* to prove that Bohr's quantum concepts can be surpassed. These exchanges mainly took place at the Solvay meetings in 1927 and 1930. The two eminent scientists had the highest regard for each other even when unable to agree. Willis Lamb expressed this to the present author by stating: "Einstein and Bohr insisted on not understanding what the other one was saying." Today it is recognized that Einstein's philosophical stance is clearly defendable; it only does not agree with the Bohrian epistemology derived from empirical experience.

The account of these discussions is beautifully reported by Bohr ([36], p. 32) and will not be covered in detail here. However, some reformulations of Bohr's earlier ideas can be regarded as significant enough to warrant presentation.

Bohr stresses here again the fact that quantum theory, in principle, only offers probabilistic laws. Their character is different in aims from the familiar applications of statistical considerations in mechanical systems of great structural complexity. The probabilistic aspect of quantum states is carried by the individual system. An observed diffraction pattern arises even if the beam is attenuated to contain less than one particle at any one time. If we have one particle only, its detection at a position X prevents instantly its detection everywhere else. Today this feature is referred to as "quantum nonlocality", but it is an unavoidable consequence of the probabilistic character of quantum predictions. Formalism does not allow any dependence on actions in regions where the particle provably has not been observed. Quantum physics does not allow us to introduce "counterfactual considerations".

Discussing the interpretation of quantum physics, Bohr points out that ancient thinkers had to exercise great caution in ascribing attributes to "Providence" using everyday language. As a case in point here I suggest the views offered by Thomas Aquinas.

Considering the probabilistic aspects of quantum physics, their origin is assigned differently by the different initiators of the theory: Dirac blames the choice on Nature, Heisenberg on the action by the observer, and Bohr advocates the experimental setup and complementarity. Whether any of these views helps to clarify the situation is not obvious.

Bohr advocates the mutually exclusive relationship between the practical use of any word and its strict definition. This reminds me of Russell's claim ([40], p. 106):

> Thus the more precision we give to a concept, the oftener it can be proved to be inapplicable, and the seldomer it can be proved to be applicable. When it is complietely precise, it can never be proved to be applicable.

The *gedanken* experiments devised by Einstein and refuted by Bohr are well known and described in many texts. Hence I omit them here and proceed with the arguments of principles. Then we come to the famous 1935 paper by Einstein, Podolsky, and Rosen. Here the quantum theory is argued to be incomplete instead of being incorrect as suggested by the Solvay discussions. The paper is written in a formal style, usually attributed to Podolsky with his solid basis in formal logic. They formulate the criterion for an "element of physical reality":

If without in any way disturbing the system we can predict with certainty the value of a physical quantity, then there must exist an element of reality corresponding to this physical quantity.

Then they devise a quantum state allowing two different measurements and claim that two complementary entities can both be assigned values, hence quantum theory is incomplete. Today one may refute this by pointing out the disallowed use of counterfactual experiments.

Bohr, however, counters by resorting to his concept of complementarity ([38], p. 73). According to him, an unambiguous meaning of "physical reality" cannot be deduced from *a priori* philosophical conceptions but must appeal to experiments and measurements. We are free to choose if we measure one quantity or the other but not both. The utilization of complementary measurements to draw conclusions means relying on experiments which differ in "the very conditions which define the predictions regarding the future behaviour". Bohr claims the phrase "in any way disturbing the system" is ambiguous. Today we would say that the situations can be unambiguously defined only if the appropriate boundary conditions (in space and time) are declared. These provide the prior information to condition the quantum predictions. They must be considered part of any phenomenon to which the term "physical reality" can be attached. Consistency is guaranteed by complementarity.

The position by Bohr has been accused of abandoning realism in favour of a positivistic idealism. This interpretation is not convincingly entailed by his exposition of the approach. Honner offers the following possible vision of a future mode of interpretation [29]:

In other words, Einstein's "locality assumption" either requires modification or we need to articulate a new notion of "quantum reality". Bohr's "higher order positivism", according to one view of reality, may turn out to be a new kind of realism in a more general account of the physical world.

Reality does not assign properties to individual objects but to "states of affairs" including the appropriate constraints. Compare this with Wittgenstein's ([10]):

A state of affairs is a combination of objects.

It is essential to objects that they should be possible constituents of states of affairs.

If I know an object, I also know all its possible occurrences in states of affairs.

Now Bohr presents one of the much-quoted declarations about our role as observers ([36], p. 62):

It was incidentally pointed out that even the psycho-physical parallelism as envisaged by Leibniz and Spinoza has obtained wider scope through the development of atomic physics, which forces us to an attitude towards the problem of explanation recalling ancient wisdom, that when searching for harmony in life one must never forget that in the drama of existence we are both actors and spectators.

This is one of the few places where Bohr refers to any classical philosophers. Here Bohr also defines a "deep truth" as one where also the opposite contains a "deep truth". No argument by Bohr is transparent and straightforward. He much admired Schiller's poem *Spruch des Konfuzius* ending in[1]

Soll sich dir die Welt gestalten,
In die Tiefe musst du steigen.
Soll sich dir das Wesen zeigen.
Nur Beharrung führt zum Ziel.
Nur die Fülle führt zur Klarheit,
Und im Abgrund wohnt die
Wahrheit.

Usually only the last two lines are quoted; the whole poem is an eulogy to epistemology.

3.2.3. The late views, 1958–1962

Philosophical résumé

In 1958 Bohr gave a lecture called "Quantum Physics and Philosophy" [37]. This gives a good summary of the philosophical ideas he ultimately settled for. He states:

In conformity with the non-pictorial character of the formalism, its physical interpretation expresses the laws of an essentially statistical type under given conditions.

The probabilistic aspects do not derive from the usual descriptions of complicated systems but are intrinsic to the theory. The logical demand of communicability requires that experiments and readings are given in "plain language" suitably "refined" by the usual physical terminology.

It must be recognized that in quantum theory we are dealing with a purely symbolic procedure. Strictly speaking, physical theories only give rules to calculate expectations about observations carried out under well-defined experimental conditions. *Operationally they only predict correlations occurring in records of measurements.* The unambiguous physical interpretation requires reference to a complete experimental arrangement. The introduction of mathematical symbols and operations is based on simple logical use of common language. Any logical contradiction is excluded by the mathematical consistency of the quantum formalism. Bohr does not accept any need for new formalisms like e.g. "multi-valued

[1] To get the shape of the world, you have to step into the deep. If you want to find the essence only zealous striving leads to the goal. Only wholeness gives clarity and in abyss dwells the truth.

logic". The word "measurement" is used in the plain meaning of a standard comparison. He, however, speculates how the theory might change in the future ([37], p. 6):

It may require still further abstractions to account for novel features, but there is no return to the pictorial representation of relations between cause and effect.

The shape of physical knowledge is not given *a priori* but is a method to order and survey human experience. This emphasizes the inseparability of knowledge and the possibility of inquiry. The role of classical descriptions in quantum physics dictates what we can communicate about, what we have done and what we have learned. It is impossible to demonstrate the existence of an independent reality, for the logical consistency of a theoretical description only implies its possibility. We progress by putting questions to nature in the form of experiments. The quantum uncertainty prevents the introduction of sharp concepts in physics. Hence Bohr, in his later writings, uses the term "indeterminity".

As reported by Aage Petersen ([42]), Bohr stated:

We are suspended in language in such a way that we cannot say what is up and what is down.

Life and knowledge

The richness of conscious life demands different placing of the section between subject and object. The concept "consciousness" is unavoidable when the behavior is so complex that its account in common language refers to "self-awareness" ([37], p. 16)

Indian thinkers gave up the question of the "meaning of life" by stressing its futility, because the word "meaning" implies comparison. And with what can we compare the whole existence?

The irreversible character of the physiological processes concerned is clearly reflected by the degree to which all that has ever entered consciousness may be remembered. As long as we speak of life, teleological terms will be used in complementing the terminology of molecular biology.

This opinion of Bohr's certainly derives from the influence of his father the physiologist Christian Bohr who in 1910 wrote ([36], p. 96):

The *a priori* assumption of the purposiveness of the organic process is, however, in itself quite natural as a heuristic principle and can, due to the extreme complication and difficult comprehension of the conditions in the organism, prove not only useful, but even indispensable for the formulation of the special problem and for the investigation and the search for a solution. But one thing is what may be conveniently used by the preliminary investigation, another what justifiably can be considered an actually achieved result. As regards the problem of the purposiveness of a given function for the maintenance of the whole organism, such a result can, as stressed above, be secured only by a demonstration in detail of the ways in which the purpose is reached.

This indicates the intellectual atmosphere in which Niels Bohr grew up. From his father's expressed views, one easily goes over to the physical approach of quantum theory. Here the unified macroscopic world, where we spend our lives, derives from the detailed mechanisms provided by the quantum occurrences. These offer unambiguously constrained physical phenomena which can be investigated as closed totalities. Which questions we may be able to answer scientifically are to a certain extent up to us ([38], p. 145):

> Under these circumstances it is, of course, to a certain degree a matter of convenience to what extent the classical aspects of the phenomena are included in the proper quantum-mechanical treatment where a distinction in principle is made between measuring instruments, the description of which must always be based on space-time pictures, and the object under investigation, about which observable predictions can in general be derived by the non-visualizable formalism.

The ensuing picture of the everyday world we live in is a possible limit but *it cannot be derived from the theoretical formalism.* It just perseveres.

3.3. Rosenfeld's approach

3.3.1. The life of Léon Rosenfeld

Niels Bohr's texts are proverbially difficult to read. On fundamental issues, his exact positions and declarations are hard to pin down. This derives partly from his limited familiarity with abstract philosophical terminology but, no doubt, mainly from his ceaseless striving to find exactly the right verbal expression for his intuitively experienced vision of human knowledge as manifested in reports on physical observations. It is therefore of interest to approach the so-called "Copenhagen interpretation" from the point of view represented by Bohr's close collaborator Léon Rosenfeld. He was well versed in the philosophical tradition and pursued a living interest in the sociological aspects of science and its teaching. He has also been reported as taking a deep interest in the Marxist approach to knowledge, even if this undertaking is mainly exhibited as a polemic against its official opposition to quantum theory as a form of idealism. The Marxist approach to physics is elucidated in the next section. However, his views are very succinctly stated and, being derived from a long intercourse with Bohr, may be taken to give a fair representation of Bohr's views as filtered through the critical scrutiny by Rosenfeld.

Léon Rosenfeld was born in 1904 and died just short of his seventieth birthday in 1974. He did his physics studies at the University of Liège, where he got his doctorate in 1926. He immediately got to interact with some of the foremost scientists of his time: de Broglie, Brillouin, and Langevin (in Paris), Bohr (in Göttingen), and Pauli (in Zurich).

From 1930, Rosenfeld taught at the University of Liège. In that year he was also invited to Bohr's institute in Copenhagen, and thus started a collaboration lasting to Bohr's death in 1962. During these three decades, Rosenfeld acted as a personal assistant to Bohr, and much of the epistemological framework of quantum theory derives from their collaboration.

After his initial stay at Copenhagen, Rosenfeld continued his teaching in Liège and subsequently worked as a professor in Utrecht from 1940 to 1947. In 1947 he was called to a chair in theoretical physics at the University of Manchester, thus contributing to the role this university played in establishing modern theory in Great Britain. In 1958 Bohr invited Rosenfeld to take up a chair at the Nordic institute for theoretical physics, NORDITA, where he was to stay to the end of his life. In addition to his scientific contributions, he also influenced the world of physicists by founding and, for many years, running the journal *Nuclear Physics*.

Rosenfeld contributed to many fields of physics. Best known is his collaboration with Bohr on the formulation of the measurement problem for electromagnetic fields. This work has remained an unchallenged milestone in the epistemological debate about quantum theory, even if its impact on the practical pursuit of experimental physics is only marginal. Rosenfeld's contributions to pure physics will not be covered here, because I want to concentrate on his explication of fundamental questions in quantum physics. During several decades, he wrote and lectured on these issues, and, as pointed out above, his presentations are often more explicit and clear than those of Bohr's.

3.3.2. Rosenfeld on epistemology

In [43] and [44], Rosenfeld defends the statistical character of quantum explanations of physical phenomena. He draws on the complementary aspects of kinetic gas theory and thermophysics to argue that any description of observations based on atomic constituents naturally introduces statistical considerations with their associated uncertainties. Even if this view has gained support by the recent advances in chaos theory, it seems somewhat arbitrary to conclude that the quantum description of atoms must needs display a fundamental complementarity. Even if it remains true that the empirically established quantum of action restricts the ability to give an old-fashioned causal description of phenomena, the intricate emergence of probability in the quantum formalism introduces something entirely novel into the epistemology of scientific knowledge. Rosenfeld's attempt at a rational justification, if his works purport to be such, is far from convincing.

Rosenfeld, however, argues that the dialectic lesson from quantum theory is entirely in line with the Marxist attitude to knowledge. He writes [44]:

> At this stage, it would seem that complementarity, vindicated from the worst misconceptions, appears in its true meaning, as a decisive step in the dialectical unfolding of our knowledge of Nature. . . .

It is a cruel irony of fate that physicists who believe themselves the guardians of the Marxist tradition should be blind to such a momentous dialectical development and oppose to it the most outworn metaphysical preconceptions....

They are thus driven to a frantic chase of determinism into an infinite sequence of deeper and deeper levels of reality: a purely imaginary construction, as ingenious and as misguided as those of the medieval scholasticism.

We should remember that this was a time when Soviet physics found it difficult to accept quantum theory; this situation was changed after Fock's visit to Copenhagen in 1957.

An essential source for Bohr's thinking is the work [45] written by Rosenfeld on the occasion of Bohr's sixtieth birthday in 1945 but published in final form only in 1961. It constitutes an eulogy by an ardent admirer and may thus in some aspects mirror more the sentiments of the writer than those of Bohr. However, its personal tone and clear exposition reveal many essential aspects of Bohr's approach to scientific knowledge:

His tremendous power of intuition into natural phenomena was served from the start by an equal mastery of all the necessary mental tools, the experimenter's turn of mind as well as the mathematician's most refined methods of analysis. To be sure, the very way in which he handles the latter, as well as his general views on mathematics, betray a bias which sharply separates his from the pure type of mathematical mind; for instance, he has never been able to take axiomatics quite seriously: such abstractions are obviously too much for his strong sense of reality, for his instinctive need to connect every human activity with some concrete aspect of nature....

The strength of Bohr's philosophy lies in the exquisite suppleness arising from its essentially pragmatic character. Any attempt to make a system of it would in fact be a betrayal of its very spirit. It does not pretend to reduce the laws of nature to a few basic principles....

For him, indeed, man as an analysing subject, always occupies the central position, and the task of science can only be one of rational description, of reducing to compendious order the data of description.

One may here mention that another master of theoretical physics, Lev Landau disliked the use of models to clarify physical issues [46]. In this aspect scientists combine features of "empiricism", "semantics", and "pragmatism". Rosenfeld writes about Bohr:

In this connection he was led to profound views on the use of language, which he tries to clear up for himself by means of a remarkable analogy with the concept of multiform function and Riemann surfaces....

But just as the gist of Riemann's conception lies in comprising all the branches of a multiform function within a single concept ("the" function), so it is an essential feature of language that there is one word only for each form of our physical activity....

It is therefore an illusion that one could avoid this fundamental ambiguity by any substructure using "new" concepts.

This shows the origin of Bohr's later concept of complementarity by exemplifying how different manifestations of a unit phenomenon can determine the concept completely. The approach seems analogous to Wittgenstein's idea that a "family resemblance" can collect seemingly different aspects under one conceptual totality. But Bohr was not propagating any consistent philosophical method.

About pragmatism as a whole, although it expresses a tendency undeniably akin to his, he is, however, non-committal, as he dislikes being classified under any "ism".

This quotation confirms that Rosenfeld recognized the phenomenological features of Bohr's thinking, but as an attitude towards his philosophical direction, the testimony is rather weak: Bohr presumably only knows James's writing through his psychology; there is no indication that he knew the properly philosophical writings of the pragmatists.

Rosenfeld on complementarity

Bohr introduced the concept of "complementarity" at the Como meeting in 1927; the ultimate written version was published in 1928. This article is presumed to dispense with the obscurity posed by the novel epistemological situation in physics. Later, he developed and extended the use of the concept beyond the initial experimental framework. However, as so often with Bohr's writings, it is not easy to extricate his meaning from the verbal intricacy of the writing. The scope and significance of this concept is presented much more lucidly and comprehensibly by Rosenfeld.

In [47], Rosenfeld justifies the introduction of complementarity as a concept logically implied by the empirical evidence. He illustrates its use by the familiar situation in thermal physics. He also presents its applicability to the phenomenon of life and also its role in the general framework of human knowledge.

Rosenfeld stresses that the complementary aspects follow from the experimental fact that physical objects may manifest both wave and particle behavior. Each physical concept can only be applied within the domain set by a well-defined experimental phenomenon. To describe the totality both aspects are needed. The waves have to be infinitely extended to have sharp wavelengths and frequencies. Thus their occurrence in space and time is totally undefined. On the other hand, in interactions, they obey the laws of energy and momentum conservation, which essentially characterize interacting particles. These two aspects of the description seem to be contradictory, and hence their use must be based on strict limits of applicability. Here the experimentally established finite value of the quantum of action sets the limits for applications. The precise formulation of this is given by the commutation relations, which serve to define which classical concepts complement each other, and these cannot be numerically fixed in the same experimental situation.

The complementary concepts are, however, not rigidly exclusive of each other, they overlap to some extent. Thus we may use wave packets, approximately localized in space and time. The formalism then puts limitations on the energy and momentum they may carry; this smears the strict application of conservation laws, and the famous uncertainty relations emerge. Rosenfeld explains:

> Reciprocal uncertainty relations are, in fact, precisely what we must expect, from a purely logical point of view, if we are, as in quantum theory, in a situation in which contradictory pictures of the phenomena have to be resorted to in a complete account. In order to avoid trivial mistakes which would arise from an indiscriminately use of such contradictory concepts, we must have some way of delimiting their domain of validity, some prescription enabling us to decide in each case how far we can go with the help of each of them. The uncertainty relations tell us, so to speak, the price we have to pay for applying to the description of phenomena two sets of concepts which are strictly contradictory. . . .
>
> It involves a new logical framework within which to arrange our conceptual frameworks, compared with which the classical way of speaking appears very narrow because it is confined to simple classical logic, governed by the principle of contradiction. Of course, the classical logic remains the foundations of any argument we use, and certainly of our mathematics.

No change of formal framework can go beyond the epistemologically set limitations of the empirically imposed requirements. Such limits will emerge naturally when we have to resort to the classical concepts to define phenomena at the atomic level:

> In fact, what the situation in quantum mechanics has helped us to realize is that science is primarily concerned with communication of experience about phenomena, in other words, with statements made by human beings, and which other human beings are able to understand. For this very reason, all such statements must make use of the concepts of classical physics, because it provides the very language describing the results of direct observation, the phenomena which we can observe with our senses either directly or with instruments to increase the sensitivity, with recording systems composed of many atoms. If then, we talk in this language about atoms which do not follow the laws of classical physics, we meet the limitations of its use expressed by relations of complementarity.

Then again, Rosenfeld draws on the analogy between the quantum complementarity and thermal physics: the description in terms of motion among the atomic elements and the thermodynamic quantities is presented. The borderline between the two modes of description is set by the particle number relative to Avogadro's number. Extensive and intensive quantities can be found, such that decreasing the fluctuations in one enhances the fluctuations in the complementary variable. The most intuitively obvious case is the duality between energy and temperature. Energy can be fixed precisely only when the system is fully isolated, whereas thermal equilibrium may be reached only through a (weak) interaction with a larger system serving as a reservoir. Then it follows that the canonical distribution is the relevant description, where the parameter of the situation can be identified with temperature.

Rosenfeld then explicates Bohr's application of complementarity to the biological systems. Their microscopic picture only provides statistical causality, but looking at the system as a living entity, one may analyze it in terms of a final causality based on the purposeful role of its construction. These aspects are not both unlimitedly valid but combine into a complementary picture of the system.

The role of the human observer is stressed as the cause of the necessary adjustment of the epistemology:

> The present description stresses the fact that at every stage the human observer intervenes in a very fundamental way by determining the very language in which the laws of Nature are formulated. The significance of the laws of Nature, from this point of view, is not that of drawing a picture of the world independent of who is looking at it, but rather of giving an analysis of the interaction between ourselves and the external world. At first sight it would seem that one thus introduces in science a subjective element. But the fear that the objectivity and rationality of the old view would be abandoned is entirely unfounded. True objectivity is simply the possibility of guaranteeing that the account of the phenomena will convey equivalent information to all observers, that it will consist of statements intelligible to all human beings. The objectivity is indeed ensured in quantum theory by the existence of transformations which allow you at will to pass from the point of view of one observer to that of another, who makes some other type of experiment. It is just the invariance of the laws of physics for such transformations that guarantee the objectivity, and nothing is changed in this respect by the advent of quantum theory.

These extracts from [47] give only a glimpse of the comprehensive and clear description the article gives of the complementary approach to the description of physical phenomena. A briefer presentation of the same material is found in the article [48].

In [49], Rosenfeld summarizes the epistemological position and indicates how it was arrived at. From Møller's tale of the Danish student, through the famous discussions with Einstein, to the eventual discussion of the measurability of electromagnetic fields, Bohr arrived at a complementary logical dialectic, which differs and surpasses that of traditional forms. It all derives from the insight that any communicable physical phenomenon must be a closed description, a final irrevocable recording must have taken place:

> Moreover a complete definition of the phenomenon must essentially contain the indication of some permanent mark left upon a recording device which is part of the apparatus; only by thus envisioning the phenomenon as a closed event, terminated by a permanent record, can we do justice to the typical wholeness of the quantum processes and stay clear of all the paradoxes to which a lax terminology may lead.

According to Rosenfeld, Bohr's epistemological success derives from his lack of philosophical prejudice:

> Untrammeled by formal schooling, guided only by the sure intuition of the investigator of nature, Bohr rediscovered the dialectical process of cognition which had long been obscured by the unilateral development of epistemology on the basis of Aristotelian logic and Platonic idealism....
>
> Even those scientists of the nineteenth century who were most inclined to idealism were impervious to the Hegelian form of dialectics; as for the Marxist dialectics, which

might have given the materialists among them a clue, it was effectively hidden from their view by an impenetrable social barrier.

Here Rosenfeld betrays his view that Bohr's concept of complementarity is a logical consequence of Marxist dialectics. This line of thought has, however, not become part of the general argumentation about the meaning of the approach. Rosenfeld, on his side, uses this approach in [50] to contrast Bohr's view of scientific knowledge as an ever-evolving structure conditioned by the progress in experimental technique with Einstein's dream of the ultimate theory dictated by the mental power of the human brain. Rosenfeld marks this view as a "somersault into idealism", thus excluding it from the set of tools allowed to dialectic epistemology.

Rosenfeld on formalism

Niels Bohr never presented any formal decriptions of the structural relationships in quantum theory. He apparently considered the heuristic presentation of his ideas as clear and convincing in their own right. He proceeds from principles that are immediately obvious from everyday experience or well established by experiment. In [51], Rosenfeld argues in any case

> That in order to exhibit more directly the link between the physical concepts and their mathematical representation, a more formal rendering of Bohr's argument should have been attempted as part of an abstract formulation of the whole theory by the axiomatization of modern logic. Such an axiomatization of a physical theory would indeed be harmless if it were properly handled, but unfortunately this is still more difficult to achieve in physics than in mathematics and in either domain the attempt generally leads not to clarification, but to confusion.

Rosenfeld ascribes this difficulty to the need for settling on "axioms" and conducting the discourse in a "metalanguage foreign to the formalism of the theory". This makes it questionable what can be achieved by such a formalization.

> However, just the proper choice of the primitive concepts and axioms and the proper use of metalanguage demand a complete mastery of the actual physical situation which it is the purpose of the axiomatic analysis to clarify, and it is therefore a fatal illusion to hope that the really deep difficulties of the situation could be skipped or in any way alleviated by any process of axiomatization. It is for this very reason that Bohr never showed any interest in the endless discussions to which the axiomatization of quantum mechanics had given rise; in his view, formal logic could only help us to check whether we had not made any trivial mistakes in our reasoning, and he rightly regarded his task as accomplished when he had exhibited in simple words the perfect adaptation of the principles of quantum mechanics to the empirical situation it purports to describe.

Rosenfeld then proceeds to discuss the formalization presented by von Neumann, which he regards as unnecessarily involving the mind of the human observer. This step should be avoided by basing the measurement description on the emergence of an indelible recording. To argue for the possibility of such an occurrence, he refers to the theory presented by Danieri, Loinger, and Prosperi in 1962. This is a model which indeed is compatible with the Copenhagen approach as formulated by Bohr. Today, however, we know that there are many such possible schemes, but none of them will ultimately resolve the enigma of transferring quantum states into unambiguous human knowledge. As a contribution about the possibilities and limitations of axiomatization, the reference [51] presents an interesting discussion.

The actual use of modern quantum theory is based on a rather involved and intricate mathematical structure. The interference of probabilities, the superposition of various states, and the nonclassical correlations implied by entanglement can only be incorporated in a linear space with correspondence rules relating its predictions to observations. However, this structure is well known and manifestly successful in describing ever-new areas of physical investigations. This story is told in a comprehensive way in the reference [52]. This writing is not remarkable in any way, and to the extent it has not become outdated by the later developments, it still gives a simple and fair description of the quantum theory.

In spite of the central role of the observer in the Copenhagen approach, Rosenfeld strongly argues against a positivistic approach to the description of physical reality. In [53] he reviews Werner Heisenberg's book *Natural Law and the Structure of Matter*. The whole review, however, becomes an eloquent but single-minded attack on Heisenberg's alleged "liquidation of dialectical materialism". The text offers amusing reading, short as it is, but it does not further the understanding of quantum mechanics as understood by either Heisenberg or Rosenfeld.

In the article [54], Rosenfeld analyzes the emergence of the concepts of classical dynamics in terms of the recent experiments about the formation of concepts in humans. Piaget and others are quoted to show how magnitude and motion become intuitive concepts. By using the possibility of mirages that deceive the eye, Rosenfeld argues that also the formal concepts, position, motion, and causality, are idealizations which have been abstracted from experience in a manner similar to the emergence of quantum theory.

The limitations imposed on the classical concepts by the quantum of action allow us to use classically incompatible aspects in a complementary way

> If one chooses to speak about "contradiction" in connection with mutual exclusion of complementary phenomena, then one may only mean a "dialectical" contradiction; in fact, one may regard the occurrence of complementary relations of quantum theory as a clear and precise example of what is implied by the general conception of "dialectical process". The quantum theory, which comprehends the complementary aspects in a logically consistent whole, represents the synthesis resolving the dialectical opposition between these aspects.

3.4. Copenhagen summarized

For easy reference, I present a summary of the main points of Bohr's view of the science of physics:

1. "Microscopic objects" exist and their properties can be investigated.
2. The "quantum theory" is a system to order our empirical material.
3. The difficulties in describing Nature derive from our dependence on language. Nature can only answer questions we can pose for it.
4. The symbols of the theory mirror no intrinsic essence of reality.
5. There is "no universal reality" restlessly described by quantum theory.
6. Abstract mathematical symbols are tools to provide knowledge only.
7. The correct description of empirical reality is the introduction of "phenomena" well described from their preparation to irreversible recordings of measurement results.
8. In order to be able to communicate the performance of an experiment, we have to describe it in terms of classical physics (suitably extended by the tools of mathematics).
9. The consistency of the process is guaranteed by the consistency of the mathematics used.
10. The possibility to use classical concepts derives from the "correspondence principle" between massive quantum objects and their classical counterparts.
11. All conceivable experiments are not simultaneously possible. This makes our concepts fuzzy and enforces indeterminacy on the interpretation of the observations.
12. Quantum theory is a complete system where all classically introduced concepts can be selected for scrutiny. "Complementarity" guarantees that no aspect is omitted from the theory.

3.5. Bohr's pragmatism

Pragmatism of the American brand has not played a major role on the European stage of modern philosophy. It has, however, influenced the development of twentieth-century philosophy of physics in a remarkable way. This is the more astonishing because there is no evidence that any of the actors of this scene has devoted himself to systematic study of the pragmatic classics. The influence seems to have been mainly indirect, based on the endeavor to define the concept of truth when it concerns the validity of scientific propositions.

Above, I pointed out that Boltzmann clearly related scientific method to the procedure required to achieve "warranted assertibility". The following excerpt from Boltzmann could have been written by James at about the same time:

> Of opinions sufficiently often confirmed by experience, we say that they are certain and that what they express belongs to our knowledge.
>
> What leads to correct deeds is true.

Still there is no evidence that Boltzmann was acquainted with James's writings. On the other hand, it is obvious that Bohr knew about the philosophy of James, even if he never referred to him in his writings. As told above, Bohr's philosophy mentor Høffding even corresponded with James. Thus even without reading any of the pragmatist texts, Bohr is very likely to know about their ideas. One may even claim that Bohr's philosophical outlook to a large extent derived from pragmatic influences.

As explained earlier, a physical measurement can give meaning to a physical concept by defining the operations required to decide if an object falls under a concept or not. The outcome of the procedure constitutes existing information about a reality persisting independently of our perception of it. In this investigation there is no absolute meaning attached to the logical term "true". Only referring to its third meaning (see Sec. 1.2.4 above), can we interpret the warranted assertibility as a substitute for true statements. All physical propositions are conditioned on insufficient evidence and consequently subject to change and corrections.

Bohr did not necessarily assign the attributes "true" and "false" to empirical statements. Some cases are not decided (ambiguous). The existence of features of reality is independent of us but not of how we choose to act. The theory of physics is no mirror image of reality but only our tool to organize and master the terrain where we move and act out our lives. Here I remind the reader of the analogy with a map: It represents reality but only by utilizing arbitrarily chosen signs.

Bohr's approach to epistemology is thus closely similar to the traditional pragmatism of James. No statement has got meaning if it "makes no difference" or implies no discernible "directions for actions". The Copenhagen interpretation is often justifiably blamed for being positivist. Historically this contains some truth, but Bohr's approach is closer to pragmatism: Instead of regarding observability as criterion for existence (*esse est percipi*), he focusses attention on what we can do. Quantum mechanically what exists is what is "preparable". This assigns the Heisenberg fuzziness of concepts not to the states but to the preparation procedure.

As the influence on Bohr has been claimed to originate in James's attitudes, it is illuminating to consider these views.

3.5.1. William James and pragmatism

Early works:

James acquired his fame originally from his comprehensive and innovative work on psychology [55]. This book seems to have been read by Bohr as well as by Wittgenstein. The book [56] defends the freedom to believe according to one's convictions. It has been pointed out that a more adequate title would have been "The right to believe". Epistemologically this seems to follow from the fact that no amount of verified knowledge can uniquely determine the transcendental structure of reality.

Also James's book [57] on personal religious experience had an influence on Wittgenstein. In spite of the strong Danish devotion to Gruntvig, Bohr did only little to indicate any concern about religion even if his writings on culture occasionally touch on such issues.

Pragmatism

In the work [58], James introduces his concept of pragmatism. He starts from the conviction that no monistic rational method can dictate the image we use to describe reality. All information of value is based on empirical methods, i.e. our human experience. The multitude of our accumulating experience provides a pluralistic view of reality. Pragmatically, all conclusions drawn from the empirical material must make a difference because of their practical consequences.

No empirical evidence can delete our personal experience of life but only add to it in accordance with new insights derived from empirically obtained results. These are not true; they simply come and persist. Their truth is of the first flavour (Sec. 1.2.4 above), and cannot be meaningfully denied. Truth is what we say about the experience. This "truth" is of the third flavour. The assertible is what has proved to be successfully applied.

The concepts of the theoretical description, James compares to coordinates or logarithms, i.e. only tools to relate features of experience's flux. This reminds us of the map model of scientific theories. True ideas are those that we can assimilate, validate, corroborate, and verify. False ideas are those that we cannot. James summarizes his argument by writing ([58], p. 117):

> *'Reality' is in general what truths have to take account of,* and the *first* part of reality from this point of view is the flux of our sensations. Sensations are forced upon us, coming we know not whence. Over their nature, order, and quantity we have as good as no control. *They* are neither true nor false; they simply *are.* It is only what we say about them, only the names we give them, our theories of their source and nature and remote relations, that may be true or not.
>
> The *second* part of reality, as something our beliefs must obediently take account of, is the *relations* that obtain between our sensations or between their copies in our minds.

James, understandably, had no knowledge of quantum theory, but the quotation could have been adopted by Bohr. Epistemologically all our usable information about reality must derive from the empirical material. Only verifiable ideas and their relations can be assigned the epithet "true". The pertaining facts are truth of the first flavour, they just are. Any conclusions from the material can be accepted only if they work. Already Boltzmann assigned Darwinian success to evolutionary surviving principles. We can accept no "idle wheels" (Wittgenstein). This truth is of the third flavour. James discusses some consequences of his views in [59], and elaborates his concept of truth in the sequel [60].

The pragmatic concept of truth has suffered from several misconceptions: It has been confused with the transcendental concept of logical truth, which can only be applied to formal systems, i.e. within consistent logical frameworks.

A second misconception is the application of pragmatic truth to moral issues; what works is good! Here most social systems hesitate to abandon the absolute rules for human behavior. In addition, whose "good" is to determine what is true? Many *Strebers* would regard their economic progress as the "good" defining truth.

As a purely epistemological principle, James's concept represents clearly "new name for old ideas", with relevance even after the revolution caused by quantum mechanics. In its focussing on the empirical material and its carried evidence, it matches Bohr's declarations. In both cases, the structure offered carries only very limited knowledge about the "ontological" implications. We can ascribe some (probably) valid features to reality; its full framework is not available to us, and it may never be.

Problems of Philosophy

A very penetrating and pleasant overview of philosophy is provided by James in [61]. Here the stress is less on pragmatic truth and more on the empiricist approach to knowledge. Many of his fundamental views are elegantly exposed here.

James started from the conviction that no logical introspection can create useful models of empirical reality. Thought implies mental images only:

> The importance attached by certain monistic writers to the fact that any chaos may become a universe by being merely named, is to me incomprehensible.

The empirical program is to investigate "reality". We thus need to define and delimit what this reality is:

> What is it to be "real"? The best definition I know is that which the pragmatist rule gives: "anything is real of which we find ourselves obliged to take account in any way." Concepts are thus as real as percepts, for we cannot live a moment without taking account of them.

Our inborn conceptual framework contains contributions from both observations and their role in our image of the ambience.

> The world we practically live in is one in which it is impossible, except by theoretic retrospection, to disentangle the contributions of intellect from those of sense.

But our totality of knowledge and beliefs is always finite or denumerable. Thus in applying logic to the formal structures, we encounter the dangers attached to unlimited use of set theory for nondenumerable assemblies:

> The same things can always be taken either collectively or distributively, can be talked of either as "all", or as "each" , or as "any". Either statement can be applied equally well to what exists in finite number; and "all that is there" will be covered both times. . . .
>
> But if we say that "all" must be fulfilled, and imagine "all" to signify a sum harvested and gathered-in, and represented by a number, we not only make a requirement utterly uncalled for by the logic of the situation, but we create puzzles and incomprehensibilities that otherwise would not exist.

And this danger is real, because however much we learn about reality, there exists a multitude of things yet to learn:

Conceptual knowledge is forever inadequate to the fullness of the reality to be known.

What we know is to be collected like a map, an abstract representation of an aid helping us to find our way in the terrain of reality:

This map of what surrounds the present, like all maps, is only a surface; its features are but abstract signs and symbols of things that themselves are concrete bits of sensible experience.

In our network of theoretical constructs, one feature we try to capture is causality; without it no predictions whatsoever are possible. But without theory, there is no causality discernible because:

It is plain that each moment of the universe must contain all the causes of which the next moment contains effects, or to put it with extreme consicision, it is plain that each moment in its totality causes the next moment.

James does, however, consider the possibility of unpredictable, irrationally random events as conceivable.

We imagine that in some respect at least the future may not be co-implied with the past, but may be really addable to it, and indeed addable in one shape or another, so that the next turn in events can at any given moment genuinely be ambiguous, i.e. possibly this, but also possibly that.

In 1911 he could have had no inkling of quantum uncertainties, but still he writes:

... concepts current in physical science have also developed mutual oppugnancies which are beginning to make physicists doubt whether such notions develop unconditional "truth".

Here we have a clear warning that the concept of "truth" may not apply to physical reality.

Radical empiricism

In [62], James elaborates and completes his description of the pragmatic epistemological framework. The essays are from different times. Here I only take up a few issues from this work.

For James the real is all experienced in its multifarious plurality. The conception of reality is construed from all experience; no doubt James counts the outcomes of scientifically planned experiments as a part of "experience".

To be radical, an empiricism must neither admit into its constructions any element that is not directly experienced, nor exclude from them any element that is directly experienced. For such a philosophy, *the relations that connect experiences must themselves be experienced relations, and any kind of relation experienced must be accounted as "real" as anything else in the system.*

The totality of accumulated human knowledge is what forms our basis for living and acting. We have to accept it until we have empirical reasons to abandon cherished beliefs or correct common misconceptions. However, the totality of our experience cannot be captured and taught, but it can grow only at its edges.

To continue thinking unchallenged is ninety-nine times out of a hundred, our practical substitute for knowing in the completed sense. As each experience runs by cognitive transitions into the next one, and we nowhere feel a collision with what we elsewhere count as truth or fact, we commit ourselves to the current as if the port were sure. We live, as it were, upon the front edge of an advancing wave-crest, and our sense of a determinate direction in falling forward is all we cover of the future of our path.

We have the will (the right) to believe in the received view.

If a novel experience, conceptual or sensible, contradicts too emphatically our pre-existent system of beliefs, in ninety-nine cases out of a hundred it is treated as false.

One unmerited feature of our consciousness is the flow of time. Consciousness is not an entity but a process:

What I do feel simply when a later moment of my experience succeeds an earlier one is that though they are two moments, the transition from one to the other is *continuous*. Continuity here is a definite sort of experience; just as definite as the *discontinuity-experience* which I find it impossible to avoid when I seek to make the transition from an experience of mine own to one of yours.

This "leap" of psychological connection may relate to Kierkegaard, but there can certainly be no foreshadowing of quantum concepts here.

3.6. Russian sensitivity

In his early writings, Bohr seemed to be careless with the interpretation of the uncertainty relations. Perhaps influenced by Heisenberg, he tended to refer to the attributes of physical objects as being created by the observation. This approach lies behind the attack by Einstein, Podolsky, and Rosen. If physical attributes are created at a distance, ordinary causality becomes paradoxical. This controversy led to an extensive clarification of how the issues stand in quantum mechanics. Especially his famous dialogue with Einstein served as a source of novel opinions on the philosophical status of the theory.

In the Soviet Union, the mental implications of Bohr's writings initiated an interpretation in terms of traditional idealism: The material world does not exist but is a figment of our imagination only. This view was in blatant contrast to the official teaching of communism. The controversy stood in the way of applications of the quantum tool to scientific development, and it was essential that this could be overcome. In February and March 1957, Wladimir Aleksandrovich Fock visited Copenhagen in order to clarify the situation. He was not only one of the foremost representatives of modern physics, but also an influential actor on the scene of Soviet physics. He managed to formulate the situation in terms acceptable to Marxism, see e.g. [63].

After this, Bohr became excessively careful with his language. Drawing behind his concepts of phenomenon and complementarity, he could express his views without being misunderstood. The problem was that he was not really understood even then.

In order to compare Rosenfeld's presentation with the Marxist one, I quote some of the philosophical classics. These texts are not usually read by Western philosophers and are mainly unknown to physicists. Already Engels put down the definition of a materialistic approach to matter and its observable phenomena [64]:

> Motion in the most general sense, conceived as the mode of existence, the inherent attribute of matter, comprehends all changes and processes occurring in the universe, from mere change of place right up to thinking.

This was emphasized when Lenin launched his pointed attack on the idealistic approach to reality [65]. In part V he writes:

> The new trend in physics regards theories only as symbols, signs, and marks for practice, i.e., it denies the existence of an objective reality independent of our mind and reflected by it. The "disappearance of matter" of which he speaks, in imitation of the modern physicists, has no relation to the epistemological distinction between materialism and idealism. Materialism and idealism differ in their respective answers to the question of the source of our knowledge and of the relation of knowledge (and of the "mental" in general) to the physical world; while the question of the structure of matter, of atoms and electrons, is a question that concerns only this "physical world." "Matter is disappearing" means that the limit within which we have hitherto known matter is vanishing and that our knowledge is penetrating deeper; For the sole "property" of matter with whose recognition philosophical materialism is bound up is the property of being an objective reality, of existing outside our mind. In order to present the question in the only correct way, that is, from the dialectical materialist standpoint, we must ask: Do electrons, ether and so on exist as objective realities outside the human mind or not? The scientists will also have to answer this question unhesitatingly; and they do invariably answer it in the affirmative, just as they unhesitatingly recognize that nature existed prior to man and prior to organic matter. Thus, the question is decided in favour of materialism, for the concept matter, as we already stated, epistemologically implies nothing but objective reality existing independently of the human mind and re-enacted by it. But dialectical materialism insists on the approximate, relative character of every scientific theory of the structure of matter and its properties; it insists on the absence of absolute boundaries in nature, on the transformation of moving matter from one state into another, which is to us apparently irreconcilable with it, and so forth. Denying the absolute character of some of the most important and basic laws, they ended in denying all objective law in nature and in declaring that a law of nature is a mere convention, "a limitation of expectation," "a logical necessity," and so forth. Insisting on the approximate and relative character of our knowledge, they ended in denying the object independent of the mind and reflected

approximately-correctly and relatively-truthfully by the mind. And so on, and so forth, without end. The "essence" of things, or "substance," is also relative; it expresses only the degree of profundity of man's knowledge of objects; and while yesterday the profundity of this knowledge did not go beyond the atom, and today does not go beyond the electron and ether, dialectical materialism insists on the temporary, relative, approximate character of all these milestones in the knowledge of nature gained by the progressing science of man. The electron is as inexhaustible as the atom, nature is infinite, but it infinitely exists. And it is this sole categorical, this sole unconditional recognition of nature's existence outside the mind and perception of man that distinguishes dialectical materialism from relativist agnosticism and idealism.

Reading this contorted text, one may get the impression that it concerns modern physics, even if it was written as early as 1909. Thus Lenin became the hero of a realistic view of the objects studied by physics. As it stands, this fits the working scientist eminently well even today.

The reception of Bohr's theories in the Soviet Union is well summarized by Loren Graham in [66] A modern rather informative and balanced materialistic view is given by Omelyanowsky [67]. In this comprehensive and detailed exposition, the dialectic approach to physics is laid out in detail; here I present the argument in abbreviated form

Dialectics is an adequate form of thinking for modern physics and science as a whole that completely corresponds to the character and constantly varying content of modern science. Theory interacts with experiment in modern physics and is not expected just to explain the phenomena of reality. It is acquiring enormous heuristic significance, its main task is becoming the search for possible new forms of phenomena and new possibilities potentially existing in nature.

Modern physics thus demonstrates, with all certainty, that cognition is not the contemplative, passive perception of nature but is an active process in which the subject of perception (the observer or researcher) plays a decisive role. From the very beginning, Soviet physicists resolutely opposed the idealistic and positivist interpretation of quantum and relativist theories that were fashionable in the West.

By isolating classes of natural phenomena, by formulating concepts and categories of various degree of generality, man pursues a goal - to catch the many-sided phenomena in the net of knowledge. But that means that the cognizing person connects up the field of experiences isolated by him into a single totality of phenomena or wholeness that appears before the mind cognizing it in its most varied aspects.

The term "reality" emerges with several meanings. The most general one is that of existence: the perceptible ball exists and the perception of the ball exists; a particle exists and a material point exists; a billiard ball exists and an ideal ball exists; an electron diffraction pattern exists, the electron diffraction exists, matter exists, and spirit exists; truth exists and error exists; they are all real in this sense of the term.

The “objectively real”, the “objective”, or the “objectively existing” means existing independently of the human mind and is reflected by it under certain circumstances. In opposite to the objective, the “subjective” means “existing in consciousness”. The subjective or spiritual (sensations and perceptions, concepts, judgements, etc.) can and does reflect the objectively real in certain conditions. The “subjective” functions in scientific theories and science as a whole which objectively reflects the real world; it therefore also figures in physics and its theories. The subjective may fail to reflect the objectively real because of errors and illusions; it appertains to man’s subjective world.

Physical reality is thus objective reality cognized in a physical theory. The symbols which form the mathematical apparatus of any physical theory do not yet form a “physical” theory. For a theory in physics really to be a physical theory, concrete measurement formulae have to be provided for the symbols of its mathematical apparatus that relate these symbols to physical objects.

The operational definition of a physical quantity thus means that it is defined by the operations needed to measure it within the limits of a particular theory. However, what has been stressed is exactly what is ignored in the corresponding arguments by operationalists (and in general by positivists), which regard operations not as a means of reflecting the objectively real in the human brain but as the real itself. A real thing is an aggregate of real properties.

Physics is thus not satisfied with individual empirical measurements; it uses them to move towards exact knowledge, generalizing the empirical material and ridding it of haphazard elements. Mathematical ideas shape the notions and principles of physics, and in modern physics they play a tremendous heuristic role on their own. The fact that measurement yields information about a quantity does not mean at all that the theory of measurements should be a special case of information theory. The state of an object is something objectively real that exists independently of whether the property of the object is recorded or not.

It is necessary in order to learn about the properties of a micro-object to use a measuring instrument that combines a device preparing the state of the micro-object and a recording device providing the data, allowing us to form an opinion about the micro-object’s properties. From this angle the measuring instrument is both a preparing device and a recording device. Measurement includes both the preparation of a state and recording in the sense above. The recording device cannot fail, in accordance with its function to be a classical object i.e. a real object such that the use of it for measurement implies the existence of conditions making it possible to neglect the quantum properties. It follows from this that the preparing device also cannot fail to be a classical object.

Measurement does not create physical properties in either classical or quantum theory. It serves cognitive and practical purposes and provides information about the object being studied in accordance with the principles of the respective theory.

We speak not simply of principles being just suggested by the experiment, but of the formation of the principles of a theory dependent on circumstances appertaining to the level of development of physics and science as a whole, including philosophy and also the state of spiritual and material culture. This is the basis for the answer to the question why a theoretical system is almost unambiguously determined by the totality of observations, although there is no logical way, to extract the theory's fundamental principles and concepts from the observations.

It seems to me that, correctly understood, the dialectic approach to physics has converged towards a form acceptable to a modern philosopher of science. This does not prove dialectical materialism as the (only) fundamental approach. It rather suggests that the use of human mental equipment forces thinking into converging frameworks. However, the limitations of our language tend to suggest differences where they may not exist.

4

EPISTEMOLOGICAL INTERLUDE

4.1. Background considerations

4.1.1. Philosophy and science

We want to build up an understanding of the natural environment, use this understanding to manipulate our living conditions and gain some image of the actual structure of the reality behind the experienced environment. As science progresses, the first task provides more and more tools to use and utilize nature for our purposes. In a purely instrumental fashion, we can manipulate natural phenomena without clarifying the foundations of our activities or contemplating the significance of what we have learned. So far no philosophical issues need to be considered.

However, once we want to constitute a foundation for our methods or claim knowledge about reality, we enter the realm of philosophy. This has always provided the stepping-stones to progress, and no theory can evolve without resting on mostly unformulated bases of presumptions. Sometimes, like at the dawn of the twentieth century, this has been explicitly acknowledged, whereas at other times, like the present one, purely instrumental views dominate.

Even the best theories can only present a description of reality interwoven with features determined by the formal tools used to carry out the research. It is here that the role of philosophy has been central; no theory of knowledge can fully explain our success in mastering nature, and no assignment of features to reality escapes the criticism of mirroring the opinions and prejudices of its proponents. Thus the eternal question prevails: what do we know as humans and about what do we know this?

The sciences proceed with their own tools and methods. Their progress uses past achievements to project the successes into the future. As an activity aiming at manipulating reality, this has been highly efficient. It seems to be irrelevant what philosophical attitude the practitioners take up; realists, idealists, pragmatists, and materialists all agree on the results of purely scientific research. What philosophy contributes does not derive from scientific progress, but it is taken to underlie the whole activity; it is the science of sciences. What this is supposed to mean has changed from time to time and from one school of thought to another.

It is not possible to survey the whole field relating to philosophy of science in this context; I have at my disposal neither time nor knowledge sufficient for that task. I do, however, want to establish a few features for an outline of a philosophical approach to the scientific endeavor. This is based on my lifelong personal experience as an active field worker in physics, and what I have grasped from others along the way. I want to formulate a web of ideas how a philosophical approach to science, physics in particular, should be constructed from the perspective of living with it.

The structure emerging is indeed a "great story"; too metaphysical to be provable and too multifarious to present a closed totality of unified experience and theory. It does, however, encompass some of the epistemic and linguistic points of the scientific activities, albeit in a heuristic and subjective way.

Many of the features I present are nonstandard from the point of established philosophy. I am well aware of that, but a full analysis of the various features considered constitutes a gargantuan task taking a lifetime or more, if it is possible at all. Many famous thinkers have, indeed, failed to reach a conclusive opinion even after a lifelong struggle. Personally, I am convinced that most of the features necessary to understand the activity and achievements of scientific researchers can be fitted into the frame I propose; if the details are fuzzy, that may be necessary to allow any progress whatsoever.

4.1.2. Words and worlds

Philosophy rests on the use of words; every advance seems to derive its novelty partly from novel use of terminology or the introduction of technical terms attempting to describe newly discovered features of the human activities. Sometimes a reader may entertain the impression that an established framework is presented as novel only in view of its propitiously invented terminology. When understanding starts to dawn on the reader, he may feel that the proposed theory strongly resembles known structures, which are but hidden behind a veil of inventive verbiage.

A shortcoming of philosophical discourse is the fact that certain words become extremely laden with the meaning ascribed to them by influential thinkers; thus the words "transcendental" and "phenomenological" point so directly to certain philosophical schools that their use outside these has become impractical.

Other words are so overburdened by various meanings that they have become almost useless in a description of a train of thought. Thus the word "world" has been utilized at many different levels of discourse: The world may be our impression of the environment, it may be the semantic structure into which our empirical knowledge is embedded, or it may indicate the totality of everything, i.e. the universe. A "possible world" is one where some feature or another differs from the one we know. But then we are told that our actually existing ambience is a "possible world"; please, list for me which features distinguish this uniquely from other "possible worlds" and I will believe you. I hope to try to use the concept of a world carefully in the present writing.

The word "existence" also offers a multitude of problems and ambiguities. If it were to have a meaning, there ought to be some feature common to the existence of the geometric circle and the coin lying in front of me on the table. In either case it communicates no meaning to claim that they do not exist, but their modes of existence carry hardly even a distant family relationship. Thus we should exercise caution also when using the word existence.

Formal logic carries as a central concept the category of "truth". Mathematical theorems are true if provable, in semantics truth can be inherited through the manipulational rules of the language and in physics theoretical predictions are called true if verified by observations. But again, the exact unique meaning of "being true" escapes elusively, and it becomes hazardous to make true statements about the truth. Especially, in building up physical arguments, my strong opinion is to avoid the assignment of truth to any statement. Here *theories* are valid, *predictions* are verified, and *laws* are applicable. However, all these results are conditional, of limited validity, and subject to known or unknown modifications introduced later. Is Newton's law true or is it not?

In all these cases, generations of philosophers have struggled to clarify the issues; the results of their efforts have been more or less convincing, but always hard to reconcile with our simple intuitive view of these seemingly fundamental concepts.

4.2. The human position

4.2.1. Animal awareness

Philosophy not only strives to illuminate the foundations of knowledge but also aims at a clarification of the essential role played by the human ego, centrally situated in the stream of sense impressions and conditioned by technical progress. No investigation of the relation between knowledge and awareness can overlook the role of consciousness in this endeavor. The image we receive is, no doubt, formed by the "I" that houses the consciousness, acquires the information received, and communicates with fellow humans. Consciousness is a central starting point of all philosophical contemplation. This was the starting point of modern philosophy as formulated by Descartes, and it is an unavoidable point of departure for founding the knowledge offered by sciences including physics.

But what is consciousness? When our physical activities cease, while awake we experience a stream of temporal progress, which propagates us forward even when no events serve to count the time. From this duration we may decide to enter more active attention, we may start recording sequences of events, we may concentrate on specific tasks, or we may decide to partake in some action by ourselves or as part of some common pursuit. We experience both the feeling for the passing of time and the volition to choose a level of activity. By exercising the faculty of memory, we may see the present as the most recent part of our history, and we may realize that, one day, we may no longer exist; our life has ceased to perpetuate itself. All this forms essential aspects of our human condition, and may, by introspection, be experienced as part of our consciousness.

But the description of consciousness fits not only humans but also large parts of the animal species. Press the worm that creeps at your feet and it wiggles. It responds instinctively to your probing, but left to itself it will actively pursue its aim to feed itself and propagate its species. To claim that it is conscious of these urges is rather far fetched, but it definitely displays more awareness of its life situation than does the weed we thoughtlessly rip out of its vegetative life.

More advanced animals have a richer pattern of behavior. Watch a cat, its deliberate movements, its voluntary appeal to humans, and its thoughtful choice of movement when creeping through the garden of your house. It clearly records the features of its surroundings, it purposefully cultivates relations to living creatures, and it makes deliberate choices of tracks and behaviors. All this has developed during its maturation, it is partly preprogrammed into its physiology and the various possible modes of action evolve in combinations with the emergence of the environment as experienced by the cat. She cannot tell us about her view of the world, and she probably carries no awareness of the finitude of her life. But can we really claim that she has no conscious sensation of her own temporal existence?

You may retort that we can give a description of our mental state, we may situate our lives in the ambient society, and we may expose plans and hopes for our future. True enough, but we also live and act in many ways governed by the same animal awareness as the cat. We cannot remember how it was to learn to walk and talk, our personalities emerge but gradually from the mists of early infancy. Many skills we possess are beyond our analysis, and we cannot tell how we perform them. Not only can we not tell what it was like to start to use a language, we cannot even describe how we are able to form grammatically correct sentences in our native language without using the rules we find necessary for learning a foreign one. Using hearing and sight, we draw conclusions about the environment, which often go far beyond the limited amount of information available. Motor activities are even further from our consciousness. Ask the tennis player how he manages to play so well, ask the trapeze artist how he can find his hold, and you will find that they cannot describe their skill, they can only carry it through.

The motor skills we have learned persist; learn to ride a bicycle in your youth, and you can still do it when old. These skills are not part of consciousness, but they can be drawn upon by conscious decisions. They are just as much part of an animal awareness as the skills of the cat. But then, the mastery of language is also a skill acquired in youth and really never lost under normal circumstances. It is a skill beyond the learning capacity of animals, but its presence and use is just the same as those of the motor skills. We may thus conclude that many of our most cherished human behaviors are just as much part of our animal awareness as the simple behaviors we share with the animals. Much of what governs us is not part of consciousness, and its mechanisms are not accessible to introspection. This is mirrored in the modern concept of the *subconscious*. All this may be part of our mental apparatus which was determined by our genes but it will still be strongly conditioned by the experiences we gather during our lives. We may not know its content, but we are not able to escape its influence on our behavior.

4.2.2. Consciousness itself

Any serious attempt to organize human knowledge and our conception of the existing reality must start from the fact that we exist and experience being as a fundamental feature of our existence. All through the ages, philosophy has tried to capture this feature as a starting point for contemplation. Plato founded his approach on the soul's immediate awareness of the world of ideas, Descartes presumed that the only fact not susceptible to doubt was his thinking, and Henri Bergson and the phenomenologists asserted that the inner sensation of the flow of time determines our existence as humans. Here we encounter one of the basic presuppositions of thinking: We cannot meaningfully deny our awareness of our own being; who would then be doing the denying?

Any systematic presentation of a conceptual system has to start by assuming the existence of something. *Ex nihilo nihil fit.* A logical system assuming nothing never reaches beyond its formal foundations. We have to define the field where our arguments play out their game. Thus the human awareness of one's own being must be a basic reference frame for all thinking. This approach even includes the possibility of solipsism, which, however, can be excluded on other grounds, *vide infra.* The problem only starts when we want to capture what the consciousness is conscious of. Below, I will try to summon up some possible proposals, but for the moment we consider pure consciousness. It seems that it has an autonomous existence; it flows from sources which we are not able to penetrate, and which reside in the depth of our animal awareness. As contended above, we admit that also animals take cognizance of their being in the world and the flow of subjective time. This we are given for free, it stems from our evolution as members of the great family of animal species. We experience this, but we have no business asking how and why. Any such statement can only be constrained by the most advanced knowledge of science; no *a priori* argument can provide information about this aspect of the consciousness.

Having admitted the awareness of consciousness, we can inquire about its observable manifestations. It seems to pose a problem if we attempt to bring the activity of consciousness into cognizance, it cannot observe itself but only experience itself. We already stated that time is an unavoidable consequence of self-awareness. But there is more. If there were not, we would not have reached even this conclusion. The animal awareness drives us to question reality; even an animal seeks activities beyond those satisfying its hunger or serve the propagation of the species. In our minds we have an urge to investigate, take advantage of, and systematize our ambience. We feel strongly that consciously we may choose between different actions; this freedom of the will is an integral part of our roles as actors in the reality of our life. We experience spontaneous drives, which steer our activities and determine our attitude towards the surrounding ambience.

The individual constitutes himself by his actions and responses in the human environment. How this is possible, and how it utilizes the abilities of humans will be discussed in the subsequent sections.

4.2.3. The sensational ambience

As animal beings we are constantly influenced by our external environment. This influence is transmitted through the conventional senses and our awareness of our body including its ordinary state of existence and its special influences like pain, dizziness, and fever. Vision and hearing combined with feedback from our motional activities give mental images and pictures of our ambient conditions. Together with smell and taste they help us classify the experiences we encounter. All this is combined with the self-experience of our bodies, which we use to find our way in the complexity of the environment and classify the impressions into separate objects and their properties.

All this cannot meaningfully be denied, because it seems to us so clearly separated from our internal experience, conscious or not. Even our observations of animal behavior indicate that they use similar criteria to steer their actions in the environment. They choose their favorite substances according to smell and taste, they move carefully in a hazardous environment, and they single out well-defined objects for play and aggression. It seems that their experience of their ambience does not greatly differ from ours. Could they describe what their environment seems to look like, they would basically agree with us in their picturing. As they cannot do this, we do not really know, but at least for higher animals they act as if this were the case. The environmental experience of an earthworm would, presumably, be different, but the language we may assume that it could speak is equally difficult to envisage. The sensational influx of experiences is something we may expect to inherit from our animal ancestry. To assume all this to be an illusion is to subscribe to a conspiracy intrigue beyond reasonable belief. We hesitate to presuppose such deviousness from our solipsist subconscious or whatever spirit may govern the life of the human self.

The sensational experience of our ambience is not based on a passive reception of sense impressions. We learn to structure reality by influences deriving from its elements and acting back on them. To hit a table corner hurts and we stumble when our foot hits a stone. Alternatively we remove an obstacle, catch an object, or cause a series of occurrences to be initiated. We can only learn to live and act by utilizing an interplay between our actions and the environmental reaction. It is an experimental fact that neither animals nor humans can learn to act rationally without mutual interaction with a consistent and stable environment. A deaf person does not hear language; hence he learns not to speak. Causality in actions and experiences gives meaning to the flow of impressions we receive from our ambience.

The child moves and expresses desires and emotions before it can give verbal expression to its feelings. Initially it acts like its animal relatives, until the gradual insight that the grown-ups' audible noises carry meaning related to the impressions experienced by the child. Development of verbal skill and purposeful actions in the environment evolve hand in hand with each other. The external experiences frame conditions, which observations are relevant in which situation, and which linguistic utterances belong to this experience.

However, the reality of experiences is unlimited and unstructured. There are no naturally distinguishable features of the experience. Smell and sight are different channels but they combine with all other experiences to constitute entities of our acquaintance. An apple is different from a flame, and they evoke different urges to act. But there are no specific sense data to list and, so to speak "check off" to determine what the "state of the world" is. No complete denumerable set of logically independent descriptions can be found; experience is unlistable and its structuring does not emerge naturally. Thus it is not well defined what could be different without affecting the rest, and hence no "possible world" description can adequately describe experience.

4.2.4. In the back of my mind

We have established that our conscious selves are bombarded by an unbroken stream of external impressions originating in the circumstances of our ambience. But out of the depths of our mind we meet another flow of impressions, consisting of memories, ideas, and emotions. These are unquestioned features of our consciousness, but they have emerged from unfathomable sources. One moment we do not take any cognizance of them, the next moment they are just there. They are often recognized as originating in our accumulated experience, we guess their origin or remember their sources. But being no part of our conscious mind they have dwelled in storage we have not been able to scrutinize. Their causal origin remains unknown, and they seem to derive from a totality of inaccessible subconscious layers of our minds. In one way, this is a trivial conclusion; it appears obvious that we cannot consciously contemplate all our memories and experiences at any one time. The consciousness takes cognition of all this but cognition is always directed towards something. No self-consciousness can be conscious of itself. *This is the intentionality of phenomenology.*

The psychological research has introduced the unseen part of our mind in its attempt to give a scientific account of our human condition. The Freudian *unconscious* seems to mainly be a storehouse for memories too painful to be kept in awareness. From their hidden sources they act on our consciousness and direct us into moods and actions we feel unable to control. No doubt, such considerations are part of truth, but in their one-sided direction into sexual experiences, they are insufficient to encompass the multifarious ways mind can be steered by features manifestly not part of consciousness. The emotional and artistic impulses of our minds also derive from sources we cannot get a comprehensive overview of. Jung's *subconscious* contains both traces of our experience as humans and images of tribal memories that have been programmed into us by our biological heritage and its development over the experience of a lifetime. As a description of the normal and innovative human, the hypothesis of a subconscious offers a useful catalog of possible sources for human behavior. Jung himself extends this to a collective network of impulses that provides similarities between various manifestations of our activities beyond what we have learned and experienced. This collective subconscious is conjectured to be a transcendental source of knowledge to be tapped by the individual directly. It remains unclear if such an entity need be postulated, or if our biological origin and common social existence suffice to explain parallels in human cultures and figments of our imagination.

Independent of the direction taken by future research to describe psychology, there is no doubt that human actions and sentiments are influenced by features of the mind which are not constantly accessible to consciousness. This certainly concerns flares of emotional feelings; origins of love and hate may not be easy to explain but they are facts of life. However, even practical problems and logical challenges may be solved outside our introspective awareness. Important decisions in our lives, new jobs, marriages, etc., are often taken when we stop weighing the pros and cons; we have to "sleep on the problem". Furthermore, anyone working on formal logical problems, mathematical or computational, has experienced how a solution may emerge suddenly as out of thin air. This, no doubt, requires a deep involvement with the problem, but the effect is not to be denied.

We thus conclude that our conscious ego, which is aware of its situation and options, is influenced by impulses from layers of mental activities beyond our imminent introspection and controlled argumentation. This adds to the stream of external influences described above, and together these impressions form our instantaneous state of mind and decide the choices of actions we settle for. When all external constraints and internal emergencies are combined, we probably have little freedom to choose our actions. Problems arise only when these drives contain conflicting circumstances; then we must ponder which direction of action is most harmonious. If then such action is not prevented by external conditions, or perhaps internal prejudices, we say that our "will is free". But now this statement is really a triviality without deep significance.

We cannot avoid duality inherent in the division between observing and observed. However, the observing conscious agent observes both impressions from the ambient environment and the inaccessible subconscious. This double input puts the experiencing ego in the middle of the Cartesian duality; inputs are received from both the body and the soul. But in forging a scientific description of some experienced reality, one must necessarily assign the roles of observer and observed. Scientific investigation describes the transfer of information from one part of our picture of the world to another. This does not imply that there exists an ontologically fixed division of the elements of human knowledge, but only that the division has to be made to allow the description of the scientific process.

This picture of a conscious ego suspended between a stream of external experiences and internal impulses may not be in agreement with any strictly logical philosophy of reality, but it seems to fit the factual experience of ordinary humans. As we shall see, it also may serve as a pattern for describing the scientific endeavor to order and perhaps to understand reality. Here we may, as indicated above, assert that all humans share a common experience of our ambience; to some extent the primitive impression of the physical environment is even presumed to be shared with the higher animals. In contrast to this, we cannot know that human individuals share the same mental experiences. As we will argue below, we may conjecture that there is a certain similarity, a rough isomorphism, between the psychological activities of separate human individuals. However, we cannot know that this isomorphism is based on an identity. We may all have our closed box of mental apparatuses that look very different, but our learning experience directs us to talk and act in the same manner as our fellow humans. We have no ways to peek into anybody else's box. We all call the same objects *red*, but we do not base this on any assumption of identity between the direct sensational experiences. In some way, we have empirical evidence for this: blind persons, orienting themselves totally without visual clues, are found to form a picture of reality which largely agrees with that of seeing persons.

4.3. Ways of representation

4.3.1. Constituting reality

Human experience consists of a bewildering stream of impressions deriving from both external sensational influences and a perpetual stream of influences from mental layers below consciousness. From this multitude of mainly unstructured experiences we constitute a surprisingly well-ordered image of experienced reality. From the structure of space and time where we move, play games, and feed ourselves, we proceed to find our place in the society of our fellow humans, for example as social, economic, and sexual personas. All this construction must be performed by our mental apparatus, consciously or unconsciously. We possess this capability, we are not taught to initiate this process; it starts by itself. Animals can do the same, but they lack the linguistic ability to benefit from the experience accumulated during lived history. Through our social interactions we acquire new skills as we develop. Like the animals, we learn to move in the terrain spontaneously, but through language, we accomplish reading and writing, not to speak of arithmetics or calculus.

At the most primitive level we achieve this by constituting entities out of bundles of impressions. Size, shape, and color combine to give material objects; the objects collect into milieus, and human groups emerge as social strata. All this is accomplished by an internal urge to structure the flow of experience by positing groups of impressions as entities. This urge is given to us, and children are amazingly competent in identifying those bundles of properties that are significant for their lives and human intercourses; they are here helped by the names provided by language either in forms of ostension or operational information, offered or expected. It is in the biological nature of man to posit a world of distinct entities.

We receive impressions, and from these we construct our physical environment with its objects impressing their properties on us. But likewise, we design a verbally oriented representation of this reality and our social encounters. All this is achieved by structuring the impressions received. No information scientist can assure us that the impulses received suffice to uniquely posit the structures constituting our selves. Thus we may be equipped with a talent to draw conclusions seemingly prematurely. But most of our conjectured structures turn out to be valid, at least in the sense that they competently steer our actions and responses in the pursuit of our lives. But we may also be deceived; if an unreal reality is conjured up by tricks, we may swallow this also. However, usually such experiences derive from some malign design to utilize our impressions beyond what they are expected to encounter. Normally we have a built-in talent to form relevant theoretical structures out of our experiences. Deviations from normality are labelled as aberrations.

The material ambience is totally built up from the objects constituted. We have no problem there. But in addition to material circumstances, we often posit abstract entities, which are not natural kinds emerging from sense impressions. Such an entity is e.g. "the social democratic party". We say that the party "is of an opinion", but we do not really observe such an entity, and it cannot be located in space like ordinary objects. Its existence is rather ephemeral and certainly different from that of the chair I occupy in writing this.

Other results of our structuring activity are mainly mental constructions and seem to lack independent existence. Examples are, among others, star configurations like the "big dipper" and "average grade" in a school. The former consists of objects of material origin, the stars, but their combination just occurred to humans by construction. The average grade is the result of a formal manipulation, and it cannot be identified by pointing at anything in nature. In these cases the entities we have posited cannot be located in reality, they cannot be reified. The controversy regarding IQs measured on groups just hinges on this: do our measured numbers mirror anything existing objectively in reality?

In classical physics this was no problem. Masses and fields could well be considered as subsisting features of reality. Here modern physics has led into a morass. Relativity taught us that our naive experience of space and time is incomplete. Quantum theory, finally, refuses to tell what it is really talking about. We know how the theory should be utilized, but we do not actually know the ontological status of its elements. Do "state functions", "probability collapse", or "photons" exist as features of reality? The various attempts to sort out such confusions form basic topics of the philosophy trying to understand modern physics. For the moment we must, however, postpone our further inquiries to the later discussions.

4.3.2. The conversing human

In the human child, awareness of the world goes together with an increasing mastery of its linguistic description. We watch and acquire an increasing conception of our place in relation to the material world and the social circle of our family. Moving gives the perception of the properties of the objects surrounding us, but gradually also the roles of our fellow humans become clearer. We expect certain services from our parents and we learn to interact socially with them and solicit their cognizance. All this is achieved through preprogrammed capacities inherited by us as human animals.

When the social urges gradually manifest themselves in language, the individual already has formed an image of the reality where he lives and acts. The language is learned as an integral part of this web of established relations to the ambience. The structure is mirrored in the emerging language, but this, in its turn, offers new tools to identify and classify the impressions offered by the environment. The individual grows to know the physical surroundings better but he also learns to appreciate and utilize the human nexus constituting the social reality offered by fellow humans.

The immediate family is the natural basis from which the contact network is grown. In spite of initial shyness, encounters with new humans are rapidly recognized and appreciated. Thus, like the physical awareness of external reality, the human ego builds its place in an ever-growing circle of social contacts. This defines the individual, his self-awareness and personality, to become an essential structure of his self. The process starts at home, grows to circles of our relatives and acquaintances, but continues essentially all through life.

Through school and professional experiences, we learn more and more about the intricate human web called society, its organs, opportunities, and demands. In grown-up life, we are just as conditioned by the accumulated social experience as we are by our biological inheritance. It is pointless to argue if genes or the environment are more important, because the two are necessary conditions and they mingle together in an intertwined unity constituting the human individual. Consciously or unconsciously we are what we are both through our inheritance and its unfolding in our life's accumulated experience.

A human without neural system would not be thinkable, but likewise it is impossible to become fully human without intercourse with others, and in this language plays an essential role. Certain sad cases show that a child, growing up with deficient human contacts, finds it extremely hard to make up for this loss later in life. We become what we are by constituting the physical self in a structure of social relations. Most of our learnings in later life and all of our interactions with society utilize language as the medium of communication. Admittedly we are biologically designed to act this way, but only by manifesting this capacity in verbal intercourse do we realize our selves. We are not only biological machines but also units for data accumulation and processing. The latter takes place steered by our memories and urges from the subconscious, but it can only be utilized and contemplated within the bounds of a linguistic framework.

Most of our accumulated experience resides in verbal entities picturing our experience of the ambience. The activities of our subconscious are not usually available to us, but when they are scrutinized through introspection, we have to use words to convey what we think we have learned. Exceptions are the purely physical reactions, acts of balance, locomotion, and sports activities. These are remembered without our being able to tell how we do it.

Thus what we can make clear to ourselves or to our interlocutor consists of verbal structures acquired during our lives. We must accept that our knowledge consists of various linguistic totalities. By reading and attending school we become heirs to the information collected during the historical development of our culture. Consequently, this has to be regarded as the ultimate source of the knowledge we have access to.

We do not learn this body of knowledge detail by detail but it is offered us in chunks of internally highly connected statements of science, history, and social structures. Utilizing our capacity to grasp networks of impressions in their totality, we build up our knowledge from complexes which we experience as making sense as they stand.

We do not spontaneously experience any potential doubt or questioning we may direct towards any part or detail of the structure learned. This happens only when we encounter a cause for doubt or disbelief. There is no gradual build-up of the piecemeal accumulation of a well-founded world picture as the positivistic approach suggests. There are no well-defined logical atoms or sense data. Knowledge is thus the totality of the received view. We are conditioned to accept what we learn. The question of truth does not enter until we find a contradiction between our received view and our experience.

Sometimes it happens that our knowledge fails to serve us. An object does not behave in the way we expect or a person deceives us. Also the economic or political system may respond against our expectations. Then we have to readjust our knowledge; often only in a few details, but sometimes in a more radical manner. If your shares do not rise, you decide that your belief in the company was misplaced; if your friend leaves you or your car breaks down unexpectedly, only a minor readjustment of your knowledge is needed.

In some cases, however, no minor adjustment is sufficient. When reality proves to differ radically from your expectations, a major readjustment may prove necessary. This happened during the last century, when our accustomed physical picture of reality broke down. New structures, relativity and quantum theory, had to be constructed from the experimental material. Now this is part of our accumulated knowledge, and we expect reality to behave in the manner dictated by the theories of modern physics.

In this way, the totality of teaching by our culture is the knowledge we have access to. Its details can be tested, and they should be tested, but in the main we expect this knowledge to be valid. In our personal lives or the development of science, readjustments may be necessary. But this only concerns the contents of the knowledge, not its essential meaning.

4.3.3. Capacity for symbolic forms

When the human structures his accumulated knowledge into systems, physical, social, or mental, the experience is laid out over a web of organization based on our linguistic capacity. It may well be futile to attempt to discern the part which is necessary, *a priori*, and which part has been constructed during the acquisition of knowledge. Both parts play an essential role, and to distribute responsibility over two necessary causes seems to be an impossible task. Our conscious view of the world is cataloged according to categories provided by our language. How the subconscious is organized is harder to ascertain; deep psychology is an attempt to establish useful categories, but so far only uncertain success has been recorded.

In spite of its superficial nature, the analogy of our human knowledge to an image is striking. An even more apt simile is the map. It gives an abstract view of the terrain, with symbolic association between the signs of the representation and recognizable features of reality. It does not give a faithful picture of its object, this would be just a duplication of reality which is not achievable. On the map, a sign for a tree or a rock does not imply that we find such an entity at the indicated position, it only suggests the character of the region. Even less real are the curves indicating the height of the terrain; the line telling that we are 100 meters over sea level is not found in nature. But where the lines are dense, we know to proceed with caution. These features are symbolic signs conveying their message about what to look for when hiking. Some features are more direct pictures of their objects; roads, rivers, and lakes are interpreted as concrete indications of these features in the terrain.

The veracity of the map is also limited by its mode of representation. The position is indicated by a network of coordinates, where the choice of labelling is rational but fully arbitrary. It is a necessary but imposed feature of the symbolic representation. In it, we may, however, discern a limitation of the representation which forces it to give a misleading impression. Namely, the mapping of our spherical earth on to a flat surface imposes limitations which cannot be overcome. Of the various projection methods each has its advantages and disadvantages; they are all imperfect.

In the same way as the map, our linguistic web of description contains both symbols for constructed entities and names for naturally occurring objects. But the representation is, by necessity, limited in its veracity and many of its internal features are dictated by the structure of language and not mirroring any feature of reality. The picture of our ambience, which we have accumulated, is flawed, but it contains sufficient information to allow us to move and act in our position in life. In this aspect, we have a picture of our environment, which to a large extent corresponds to the picture a map offers of the terrain.

In the process of building up the mental view of reality and its structuring according to human social networks, we acquire the ability to represent complicated experienced totalities as symbolic webs where the concepts are related to each other and our observations via the properties offered by our ability to construct tools for symbolic descriptions. The capacity to utilize symbolic networks has been necessary to form the human individual with his relations to the physical ambience and social environment. However, once acquired, this ability seems to take on a life of its own. Man excels in building up structures not directly motivated by his experience. Art and mythology are prime examples of the results of this activity. Its highest manifestations are, no doubt, the construction of abstract mathematical systems.

The origin of geometry may well reside in the need to measure land as property and object of agriculture. But already the ancient Greek vision of geometry as a web of interrelated concepts was a remarkable feat of the human mind. Postulating certain relations between primitive entities, one could show that certain other relations follow by necessity. From this origin, mathematics has evolved into an integral part of human knowledge, and its independence of the experienced reality has become more and more evident. The realm of mathematical concepts has an existence which remarkably differs from the existence of objects we discern in the physical reality. According to customary definitions, this realm is usually labelled Platonic. Its reality has been fiercely debated, but if we realize that its concept of "existence" differs profoundly from the corresponding concept in human experience, there is neither any danger nor any inconsistency in assuming that it can be considered real. It may not, however, be regarded as given *a priori*, because the human creativity has restructured and expanded it in unforeseen and largely optional directions.

However, when humans have acquired this ability to construct rational but optional abstract networks, this activity has become an essential part of behavior. We construct social and economical structures, and also legal and political machines. These are supposed to organize our lives in human societies, but other constructs serve mainly purposes of entertainment and leisure activities. Games and sports proceed according to rules which are totally arbitrary constructions. The only requirement is internal consistency, which ties the activity to the same conditions as our efforts to extend the realm of mathematics. In this sense, mathematical activity may well be classified as a sort of game. However, the wide applicability of mathematics demonstrates the utility of these games.

Once humans acquired the ability to represent experience in the form of symbolic models, they made a habitual tool of this ability. Patterned on the use of language, researchers into the structure of reality started to employ formal models to describe the accumulated evidence from the environment. Already, the Greeks tried to apply their knowledge of mathematics to the structure of reality; even Kepler tried to justify the dimensions of the solar system by referring to the five Platonic bodies. This attempt was misguided, but its conception was no less rational than the modern trend to tie the properties of elementary particles to the possible group structures of mathematics.

The application of mathematics to a rational description of our ambient reality has proved successful beyond all expectations. Formalism can well be considered as extending our ordinary language by the tools and concepts of mathematical structures. This has been definitely stated by Bohr. Ultimately, our modern physical conception of reality has been an astonishing achievement beyond all comprehension. We have no intuitive vision of the foundations of relativity and quantum theory, nevertheless, a rational and consistent application of these symbolic structures has, so far, allowed both successful predictions and creative directions for design in the empirical reality of physics. If the ultramodern constructions of the reality from "multi-dimensional strings" prove successful, the symbolic description of reality is still further removed from a naive everyday picture of our ambience. These models may have their existence in some transcendental Platonic heaven, but their utility in guiding our actions in the physical reality has been proved beyond reasonable doubt. Just like the map, the pictorial representation serves its purpose to guide our path through the bewildering multitudes of reality.

I do not know who originally suggested the map as an analog to our symbolic description of reality. It is known and appreciated by many scientists. In 1911 it was used by William James [61]:

> This map of what surrounds the present, like all maps, is only a surface, its features are but abstract signs and symbols of things that themselves are concrete bits of sensible experience.

However, in all its simplicity, it offers an eminent simile to describe the utility of man-made rational structures in human life. It mixes features of reality with optional creations of the human mind, but it offers an unfailing tool for navigating through the entities encountered on our travel through the terrain. A philosopher might complain that it concocts ontology with epistemology in such an intertwined totality that separation of its elements becomes impossible. As a representation, it necessarily contains the limitations of the method of mapping, but as a summary of experienced totality, it offers at each moment the best picture achievable by the human ability to construct symbolic webs of description. Why the mathematical constructs can be so successful, is a metaphysical question to which it may well be beyond our capacity to find the answer.

4.3.4. The use of formal structures

I already stated that philosophical discourse is to be logically independent of the results of scientific inquiry. This may be so, but the activity of philosophers has never been independent of the state and progress of science. How could it be, if its task is to illuminate the status of our knowledge? It must at least accept what that knowledge is: what constitutes the allowed collection of models? Here the progress of science sets the boundary conditions on the methods and concepts at the disposal of the philosopher. The atoms of the Greek are not the same atoms as those of the quantum physicist, and the space and time of Galilei are not the same as those of the modern cosmologist. The mind of Descartes differs substantially from the mind of a modern Jungian. Many of the philosophical questions are eternal, but the facts offered to the analysis are new and the scope of acceptable explanations evolves and transforms. One possible explanation why philosophy has never reached the stable progression of natural sciences may well be that its problems linger on but the potential answers keep changing, being just out of reach for the final conclusive solution to the enigmas.

If Newton's law is valid, what is the mode of existence of this law? Modern physics is even more difficult to capture in terms of comprehensible mental images. Physics needs to allow the use of metaphysical concepts.

Finally I wish to address the question of unity of knowledge. We know that the sciences of sociology, economy, psychology, biology, and physics use very different methods, models, and criteria for validity. We also know that even some parts of physics use incompatible constructs for their conclusions. Quantum theory has not yet found a place in the world described by Einstein's theory of relativity. Fitting has been performed successfully at the edges of the validity of the theories, but no working merger has been found.

Are we to assume that all observable manifestations of reality eventually will fall within the realm of a single fully unified theory? For the moment we have no clue to an answer to this question. If such a theory exists, it will certainly be very far from our present knowledge. Its possibility seems highly unlikely because of the different and untranslatable concepts utilized by the different sciences. But even if there exists a unified theory of reality, residing in some Platonic universe of ideas, it may well be beyond the capacity of the human mind to find it. It may require a theoretical construct outside of our accessible store of models. If this is so, we may ultimately have to be satisfied with our patch-wise success in describing reality. But this conclusion is not yet compulsory; the physicists can still continue their quest for a "theory of everything". But today, this goal does not even encompass an understanding of the workings of the human mind. That target is hardly visible at the horizon.

4.4. Role of formalism

4.4.1. Symbols used and misused

Our conclusion has been that we orient ourselves in the environment in accordance with maps which have arisen by our application of an inherent symbol manipulating capacity to the multitude of sensational experience. This process utilizes the inherited capacity to create formal structures combined with an individual collection of life experience. How this capacity was acquired is a matter of evolutionary research, and the individual inherits the collected wisdom through the mediation of language. The success of the human species as such and its mastery of nature through science bear sufficient witness to the purposefulness of the end product. That its very success may lead to our total destruction, is a different aspect of our existence.

As with many other human activities, the utilization of symbolic representation has continued to new realms as far as intuition, imagination, and creativity can take it. This has been one of the driving forces in the emergence of mathematics. The development has been constitutive for critical thinking both within mathematics and philosophy during the twentieth century. Its origin may be considered an internal problem for mathematics, but because of the essential use of mathematical models to describe the empirical material, physicists also have to worry about these problems in the realm of pure mathematics.

The first realization, already in the nineteenth century, was the insight that different geometrical systems can be equally valid. The *a priori* status of Euclidean geometry awarded it by Kant was shaken by the introduction of alternative axioms. This should have been manifest earlier, if the autonomous status of axiomatic systems had been recognized. The insight, once assimilated, removed geometry from the realm of natural sciences and assigned to it a proper position as a free creation of the human mind. Its structure could be construed solely on the bases of logic and consistency.

This awoke the hope that all of mathematics could be built upon the basis of a logistic approach. The movement, initiated by Frege and Russell, did a lot to clarify the status of *mathematical truth* even if the original goals could not be reached. Russell recognized early that entities could as easily be classified according to certain properties assigned to them, or according to their membership in certain sets. The scope of sets had been recognized by Cantor, who investigated the magnitude of sets, their subsets, and the sets to be formed from other sets. The framework was there, and the early thinkers assumed that any set could be accepted, and no additional similarity between the members need be posited. Thus the set

{the number 5, my neighbor's car, smallpox, the social democratic party}

may not be very useful but nevertheless fully acceptable. This approach, however, turned out to lead to paradoxes usually referred to by Russell's name. I do not want to enter this discussion here, and even less the artifacts introduced to overcome the problem. In the following, I will refer to this problem as the "problem of membership".

An example of a dubious set may be chosen as follows: Take the real numbers in the interval [0,1]. Their representation as a set could be by decimal numbers of the form $0.\mathrm{b}_1\mathrm{b}_2\mathrm{b}_3\ldots$. The set can thus be formally defined as

$$S = \{0.b_1 b_2 b_3 \ldots \mid \forall b_i \in \{0, 1, , 2, \ldots 9\}\}.$$

We now form a different set by adding one requirement; the digit "9" is excluded from all numbers in the expansion:

$$S' = \{0.b_1b_2b_3 \ldots \mid \forall b_i; \exists i \in \{0, 1, , 2, \ldots 8\}, \}.$$

This set seems to be well defined, it is clearly a subset of the previous one. It can also be seen to be of the same cardinality, because it may be read as a representation of the reals in the interval $[0, 1]$ in the system of base 9 instead of the customary 10. For an infinite set this should, of course, not be a problem. But what about the set removed:

$$S'' = \{0.b_1b_2b_3 \ldots \mid \exists i; b_i = 9\}.$$

One may justifiably question if this set is well defined. A constructionist theoretician would definitely deny this. Its formation would require an investigation of all reals in the interval as expressed in decimal notation. It is doubtful whether such a process is an allowed mathematical procedure. This set does not serve any useful purpose, but its existence seems to depend on an operation which is not necessarily acceptable in the formalism.

The formalist approach to mathematics chooses to regard the mathematical activity as a game of *formal manipulations* of symbols. The primitive elements of the game are only given by the rules governing their possible roles in the game. They can be combined and related by rules laid down in the catalog of possible manipulations. This allows one to develop a formal theory of proofs. Those theorems are valid which turn up as the end products of a series of operations applied to the initially posited starting point transformed according to allowed operations. This is a rewarding activity, as far as it goes. But also this suggests more ambitious goals to be set up. Given a well-formed expression in the formal structure, is there an algorithmic method to decide if this can be proved or not? Can we make a computer that will systematically decide the validity of all possible well- formed propositions? This question was answered negatively by Gödel, Turing, and Church, each in his own way.

The result is simply formulated: No algorithm will ever produce all possible provable sentences however much we try to extend the series of allowed "proofs". The seriousness of the situation is emphasized by the accompanying results:

1. By reasoning extending beyond the original formal structure, one may conclude that some of these unprovable propositions, in fact, do hold.
2. On the other hand, it follows that no procedure may be offered that would report if a given set of algorithmic manipulations actually will serve to tell if the program applying this set will end in a proved formal result or not.

These conclusions are confusing but internally consistent. Our ability to know even the range of valid entailments from our man-made formal structures is highly limited. The formal proof is captured by the oversimplified statements

S = {A is valid}

A = {S can not be proved}

If A holds, we conclude that we cannot prove it to be valid. On the other hand, if A is not true, we conclude that S can be proved and conclude that we have proved the validity of A. But this is in conflict with our original assumption about A, and makes this assumption inconsistent. Thus we may decide that A holds but we cannot prove it. This toy-argument is necessarily a parody of the much more elaborate argument by Gödel. However, it exemplifies the quandaries we may encounter in blind applications of logical concepts to formal structures. I will call this the "problem of provability".

Thus we may conclude: Applying the formal models of mathematics to the product of our empirical activities, the extension of common language has allowed us to construct a theoretical machinery which has been tremendously successful. It has allowed us to describe the outcomes of experiments ranging from the world of atoms and elementary particles to the totality of cosmos and its intense birth in the "big bang". Its predictions have proved to be valid descriptions of coming events and their accompanying measurement results. The rules have even allowed us to manipulate reality and construct devices based on our best understanding of the foundations of the behavior of matter. Lasers and semiconductors form fundamental building blocks of modern technology. Still, without an understanding of quantum theory, we would not have been able to design such systems and manufacture them.

But all this success is built on the foundation of mathematical structures. And we have already concluded that their foundations are shaky to say the least. We have no guarantees that the structures employed are consistent nor that we may efficiently derive their implications. This situation which upset a whole generation of philosophers, has not shaken the practising physicists to any extent. On the contrary, spurred by the success of modern physics, they feel free to suggest any formal explanation for observation extending the knowledge of experimental outcomes into new regions. The issues of validity and consistency are comfortably deferred to the future. Following Popper, they expect that invalid theories will be weaned out by disagreements with experience, and they worry about inconsistencies only when they are encountered. And they are often taken care of by *ad hoc* prescriptions which remove the symptoms but do not go to the roots of the problems. A case in point is the so-called "renormalization theory of fields". Here the formalism developed avoids facing the origin of the difficulties but devises a workable operational scheme which, however, has turned out to be highly successful in the whole realm of applications of field theory.

One may justifiably ask what is characteristic of the physicist's activity, which allows him this leeway of action, where all doubts and paradoxes may be swept under the carpet?

4.4.2. Tests and classifications

How do we apply philosophical considerations to a natural science like physics? Much has been written about this, but little conclusive illumination has been gained. I claim that the reason derives from philosophers' quest for a certainty that cannot be achieved in symbolic descriptions of reality. All formal structures pertaining to empirical cognition, are only temporary and tentative. It is the best guess we have achieved and subject to both known limitations and future modifications. The concept of truth can only be applied within formal structures, not to descriptions of reality. In fact, there are no 'Laws of Nature'. The marvel is that, nevertheless, mathematics works so efficiently in describing the empirical cognition of reality.

One problem with the formal descriptions of sciences and, in particular, of physics is their static character. Science is primarily an activity; we collect observations, apply operations to systems and observe their responses, and finally posit classifications of the observations in terms of some internally consistent symbolic framework coherent with the collected data sets. To what extent we assign the terms and processes of the theory to features of reality, is often a matter of metaphysical convictions or pragmatic outlooks.

A philosophy of physics must necessarily contain the concept of an operation performed on the domain of the part of physical reality we have chosen to investigate; we must be able to describe experiments and their outcomes. To this end, we assert as a primitive concept, the operations which query whether a given concept is applicable to a selected entity or not. A concept which cannot be so introduced may play no role in our description of nature. This is operationalism as a strategy, not as metaphysics. Our theories may well contain many entities, which we cannot subject to such tests. Whether they are supposed to have counterparts in reality is often a moot point: As an example take the wave function. It is objective in the sense that all its users agree on its predictions and applicability, but its mode of existence is still unsettled. A pure positivist should deny existence to such "theoretical terms", whereas a pragmatist would easily reify all such entities which play a role in the structure of the theory. Both may, however, be partly right, but ultimately this may be more a feature of mentality than reality.

The physicist decides which possible tests his theoretical description entails, and applying these he obtains various answers from the experience. These are often real numbers but, for simplicity, we concentrate on simple binary questions. This is the situation often utilized in quantum theory, and corresponds well to the information theoretic consideration of knowledge. The numbers emerging can easily be fitted into this framework, if we consider experimental uncertainties. The question may well be: "Is this value in the range $[x, x + dx]$ or not?"

When we have performed a set of such measurements we collect the results into bins according to outcomes. Each such bin then contains an assembly of observations where the outcome of the operation is the same. We have classified all the observations according to the result of the experiment. We may thus assign the physical property we tested for to all situations ending up in the same bin. We can thus consider the property as a theoretical construct in our symbolic description or alternatively as the assembly of such situations where the corresponding operation gives an affirmative answer. This dual consideration is well known from logical argumentations, and its application to empirical science is not controversial. To what extent these assemblies are assumed to be existing elements of physical reality is necessarily determined from the status of the theory and the user's attitude to empirical methods. No universal rule can be applied here. The concept of weight of individual pupils in a class is, presumably, an objectively construed entity, but the average weight has no corresponding element in reality.

In some cases we may find that certain concepts go together. The same events show a tendency to bunch in the same assemblies of outcomes. This suggests a correlation between them, but without theoretical analysis, we must not assign a necessary connection between them. This is the problem of Hume: no observed connection is, in itself, sufficient to ascertain a causal relationship between them. To introduce "causality", we must have a theoretical framework, where the symbolic concept associated with one feature of reality formally implies the validity of another symbolic concept associated with a different physical feature. Then there can be assigned a causal connection between the two phenomena, but its necessity resides entirely within the theoretical framework; no *necessary causal connections* can be assigned to the phenomena themselves. The only necessity binding reality is that "it is what it is"; no further features can be proved from necessary, purely logical considerations.

We classify phenomena observed in probing reality by identifiable operations according to the symbolic structures we set up against the multifarious possibilities to act by the scientist. The structures are not derived and cannot be so derived, from the empirical material. In simple cases it may suggest principles and relations, but the amalgamation of empirical outcomes with a theoretical structure is a truly creative activity which cannot be carried out bit by bit, but must rest on identification of formal structures with observed trends in their totality. The model relationships chosen are strongly conditioned by the mental ability to construct formal structures, but the empirical facts impose other strict requirements in selecting a propitious theoretical construct. We are, in fact, far from any *a priori* categories imposing their pattern on the scientific description of reality.

The theories of observed reality are thus consequences of the mind applying its remarkable ability to assign webs of logical constructs to the accumulated results of the physicist's probing and observing physical systems under varying conditions. The emerging theoretical picture is thus conditioned by the unambiguous description of the action taken to obtain a result. Only such a full description can be associated with a phenomenon, just as repeatedly stressed by Bohr when quantum theory was discussed. To what extent independent existence should be assigned to the various individual elements needed to carry out the logical analysis, remains a point of controversy and interpretation. To make the endeavor work we need not take an absolute stand on this issue.

4.4.3. Overcoming paradoxes

I have outlined a description of the scientific method to obtain symbolic descriptions of observed reality. It contains no assignment of truth to statements beyond the assertion that observed data are obtained using valid and agreed methods of operations on physical reality. No necessary causal connections exist between recorded features of observations; they can only be causally connected through some agreed-upon rule of the symbolic framework to which the observations are associated. The phenomena, including an unambiguous prescription of the method to obtain information, require a nexus of concepts. The individual components of the narrative may not require an independent explication or even existence.

A property is an answer to a yes-no question applied to an empirical situation. A phenomenon or an entity can be said to have a property if it is classified with other similar situations when the outcome of the proper test gives an affirmative answer. This is not new, it is the old method of introducing concepts by equivalence classes. Here, however, the difference is that we do not classify entities according to properties they possess, but according to the outcome of performed operations that we have agreed test for the assignment of this property. To ascribe the property where it has not been tested, constitutes a counterfactual statement, which should not be applied in quantum physics.

More complex relations can be handled in exactly the same way. Pairs of elements of the observational reality are found to belong to a class of pairs reacting equally to tests designated to probe just this relationship. To have meaning, this test must, of course, play a role in our symbolic description of the relevant physical situations. A most prominent relation of this type is, of course, causality: If "A then necessarily B", where the "necessity" belongs only to the formal part of our description. If the sun does not come up tomorrow, this would of course be bad for our living conditions, but it would be a fatal blow to most of our theoretical understanding of reality.

A property is thus found to be represented by a set of such situations where a certain test is successful. But now we must notice that such a set always has a finite number of instances. No experimental fact rests on an infinity of observations. At most, we may be able to extrapolate our investigations to denumerable sets of instances. I call a set *numbered* if it consists of a finite or at most denumerable assembly of individual instances. All sets occurring in science are of this type.

However, as the procedure we use to construct the sets, we assign to them a property which can be tested by the science in question. This follows from the choice of operations as primitive concepts necessary to make sense of the scientific activity. But this solves the problem of membership; if an entity is said to belong to a given set, we know that it has passed the correlated test. This thus holds for any member we pick from the set. There is no necessity to make statements about all the members of the set. We only need to know what holds for those we behold.

The symbolic structures we accept as valid descriptions of reality contain certain logical constructs and rules for assigning validity to statements. The rules making these correspond to operations in reality imply certain conclusions about the configurations experienced. But these statements are derived from numbered series of observations and they can thus be tested only on numbered instances of observations. The logical conclusions taken within the formal structures are relevant for reality only as relationships between numbered occurrences. Because any theory, as such, can be represented by numbered models, we can achieve all useful conclusions within the formalism of sets with numbered memberships. This is a computable realm, where all that needs to be done can be done, and no paradoxes are expected to occur. Any problem we encounter has to be based on numbered inputs and any conclusion should derive from arguments finitely provable in the formalism. We can only know a numbered amount of statements about the outcome of a theory applied to our description of reality. Any statement to the contrary would imply that our description of reality contains features we may not be able to *simulate on a computer*. We cannot prove this to be possible, but if some theoretical argument suggests it, I would be inclined to think that the symbolic formalism is badly construed rather than that reality demands it. We may, of course, use formal constructs that are larger than numerable, space and time may justifiably be considered continuous, but any series of events placed in space-time constitute a numbered entity. This seems to be a way to avoid the "provability" problem when mathematics is applied to physics. The statements of observational facts are mirrored into numbered sets, and no valid conclusions can be found solely in the symbolic network of theoretical relationships. Assume that a statement, formulated in an unambiguous way, is seen not to be provable in our theory, but it is found to be supported by the experiments. We would not then conclude that the statement is unprovable but true in the Gödel sense; that would be a ridiculous attitude for a scientist. We would rather look for an extension or amendment of the theory to accommodate the validity of the observation. Thus new experience is included and assimilated into the theoretical framework of science. No formal paradox is allowed to act in the procedure of theory construction, and controversial subtleties are avoided by the method itself.

The attitude above requires no major alterations in the formal logical procedures. I contend that a set is made up by elements where the property A has been verified for all its actual members. Picking an arbitrary element out of this, we know that it has property A and conclude that it satisfies all the predictions our theory tells us are implied by A. We are, however, not allowed to state that "all the elements that have property A, must necessarily comply with all the consequences stated by our theory". This is often done in mathematical discourse as part of the logical process of conclusions. The equivalence is, however, not a necessity. The problem is that we should not quantify over elements we have not yet taken cognizance of. In physics this care is obligatory. Any statements we make about reality, will always rest on a numbered basis, and may prove to be subject to adjustments. This potential source of inconsistencies was noted by William James already in 1911 [61]:

The same things can always be taken either collectively or distributively, can be talked of either as "all", or as "each" , or as "any". Either statement can be applied equally well to what exists in finite number; and "all that is there" will be covered both times....

But if we say that "all" must be fulfilled, and imagine "all" to signify a sum harvested and gathered-in, and represented by a number, we not only make a requirement utterly uncalled for by the logic of the situation, but we create puzzles and incomprehensibilities that otherwise would not exist.

The mathematician Hermann Weyl was strongly influenced by the intuitive theory of Egbertus Brouwer. In addition to the use of the excluded third, the movement was strongly critical of conventional set theory. Weyl writes about criteria used to build sets of numbers [68]:

The theory of sets unhesitatingly makes use of such criteria which it sets up to decide whether a point or a number belongs to the set or not. One can see that it thereby becomes involved in fatal logical circles.

In the next section, I will try to present a formal structure realizing the ideas expressed up to now. This will rely on more technical tools of abstract methods, and requires a familiarity with the notation of logic. Anyone who cannot enjoy the attempted presentation may, without loss of continuity in the argumentation, skip this part and jump to the following section.

4.5. Formal procedures in physics

4.5.1. The test (forming a judgement)

Weyl describes the method by which scientific concepts are formed [68]:

We have found that elementary analysis must be carried so far as to establish elements varying each exclusively within a range of possibilities which can be surveyed from the start; because it originates from free construction. This is one side of the matter which I designated by the term constructive generation of a field of variation. The other side which here mainly interests us is the fact that the subsumption of the particular concrete case, the "individual", into this field does not take place on the basis of immediately recognizable characteristics, but is the result of mental or physical manipulations or actions to be performed on it. To determine number, for example, one has to apply the process of counting; to determine the mass of a body, one has to allow it to interact with other bodies.

Formally considered, all scientific results derive from the fact that we may test for the presence or absence of a property Ω, observable in nature. If it is not observable, it cannot be scientifically investigated. In order to make sense, the investigation must also have available objects to which the test may meaningfully be applied. If there is nothing to explore, no scientific procedure makes sense. The measurement consists in assaying a given entity against a standard, the result being a real number or a set of such numbers. To avoid the problems attached to the existence of an empirical continuum, we assert that all such tests give a yes-no result. This binary description of reality agrees with the decree by Wittgenstein

Each item can be the case or not the case while everything else remains the same. (Tractatus 1.21)

The binary description of the empirical reality is characteristic of the early Wittgenstein, but it also recurs in the presently topical quantum information interpretation.

If we test the entity A for Ω and find an answer, we denote this with

$$\Omega\,(A) = \omega,$$

where

$$\omega \in \{yes, no\}$$

If the test has been carried out and given a "yes" we write

$$\mho A \Leftrightarrow \Omega\,(A) = yes.$$

This introduces the new logical operator $\mho$ which resembles the Russell notation $\vdash$ but here it is introduced to assert the fact that an empirical question has received an affirmative answer. The operator $\mho$ is a kind of modal character. If the property is not verified, it may be false or not applicable. Any proposition containing it can only be entailed by other propositions containing this operator. It may be used to introduce empirical conclusions into a logical argument. An example

$$\text{Charged (Electron)?} = \text{yes}$$

is equivalent with

$$\mho_{\text{charge}}\,(\text{electron})\,,$$

which demonstrates how we may indicate the property to be tested for.

When our empirical procedure has brought forward an assembly of objects: $A_1, A_2, A_3 \ldots$ we may collect the ones giving $\omega = yes$ (and only these) into a set

$$\mathcal{A} = \{A \mid \mho A\}\,.$$

This set is obviously finite (or at most denumerable). It is thus the set constituting the range of applicability of the concept Ω. Once this set has been obtained it may be used in the ordinary way common in naive set theory. No anomalies are possible because of the construction used. The normal logical operations on this set are valid. The elements $\{A\}$ constitute our model of the empirical material. Their relevant manipulations are determined as the theory we hypothesize for the field of observations.

The test Ω is in some way the equivalent to the "meaning" (*Sinn*) of the physical concept; the corresponding "extension" (*Bedeutung*) is the ensuing set $\mathcal{A}$. The strict logical investigation of $\mho$ as an epistemic modality will not be carried out here.

Once we have constituted the set $\mathcal{A}$ we may use it to carry out investigations on it:

1. Confirm the result of tests of A_i deciding $\mho A_i$.
2. Structure the physical entities contained therein by introducing further experiments (tests).

Let us assume we have found a different property tested by Ω_b defining the set

$$\mathcal{B} = \{B \mid \mho_b B; B \in \mathcal{A}\} \tag{1}$$

This defines an additional physical property carried by the elements of $\mathcal{A}$. We may also form the complement

$$\bar{\mathcal{B}} \equiv \{B \mid \neg\mho_b B; B \in \mathcal{A}\}\,. \tag{2}$$

This is the set where the property Ω_a holds but the property Ω_b does not hold. This may include the case when this property is not applicable.

Example: the set of all bosons structured according to their mass. Here, for no value of the mass m do we get a "yes" for the photon.

Now we encounter the cases:

- All elements B of $\mathcal{A}$ either have the property Ω_b, or if not, they satisfy $\neg\mho_b B$.Then we may write,

  $$\mathcal{A} = \mathcal{B} \cup \bar{\mathcal{B}}.$$

- If this is not the case but we may find a complete set of mutually exclusive properties $\{B_i\}$ such that

  $$\mathcal{A} = \cup_i \mathcal{B}_i$$

  and

  $$(\forall A \in \mathcal{A})\,;(\exists i)\,;\mho_{b_i} A.$$

  This is a complete classification of the set $\mathcal{A}$ distributing a collection of physical properties over its members. The symbolic representation states that every A belongs to, at least, one subclass $\mathcal{B}_i$
- Starting from a physically accessible set of properties $\{B_i\}$, we may find that they do not effect such a complete classification. Then there is a remainder

  $$\mathcal{R} \equiv \{R \in \mathcal{A} \mid \neg\,[(\exists \mathcal{B}_i)\,; R \in \mathcal{B}_i]\}\,.$$

The concepts introduced by $\{B_i\}$ are then not universally applicable to the elements collected by $\mho_a A$. We then have

$$\mathcal{A} = (\cup_i \mathcal{B}_i) \cup \mathcal{R}.$$

4.5.2. What entities can we posit?

In order for a physical property Ω to be awarded existence it has to be consistent enough to allow unambiguous communication of the outcome of the corresponding observation, the judgement. This implies at least the following demands:

- The property tested must possess enough durability to be identified by various observers and stay accessible for repeated testing by all such potential investigators. This does not mean that the outcome of the test stays always the same; nay it may vary with the circumstances of the observer or in the course of time. However, there must be enough persistence to allow us to discern the identity of the concept from one observer to another and with time. This may require some hypothesis drawn from our theoretical modelling of the field of observations. An example is "velocity", which depends on the movement of the experimenter and the time of observation.
- The concept must allow for an empirical investigation, and to have potential existence it must be accessible from several points of view. It must be consistent within the theory being constructed. All methods may not be applicable to all cases, but the various methods must be applicable in overlapping domains to allow the extension of its meaning to the totality of all possible instances of observation. The distance to the distant galaxies is not *a priori* an entity of the same character as Bohr's radius. The concepts are then forming systems reminding us of the mathematical construct of charts of overlapping maps. A concept applicable only once to a singular situation is of no use in science: We do not regard the number of clouds as a useful physical observable. To connect the various aspects of physical concept may require the imposing of a theoretical structure on the empirically obtained material. This allows for Bohr's concept of complementarity.
- As experimental outcomes, the results of observations will be affected by accuracy limitations incurred during the observational process. Such are the unavoidable errors afflicting any real observations. These must, however, be stable in occurrence and consistently mirror the imperfection of the technique of observation. Within these imperfections, the different possible tools of measurement should be compatible. When technology is developing and improving, the errors should decrease in magnitude. When consensus persists among the various investigators, they are permitted to posit the existence of an entity corresponding to the observed quantity. Its absolute existence is not ascertained, nay it can never be made certain, but it may still play a role in the development of physical hypotheses.

Only concepts or properties so introduced may form part of a consistent and useful theory of empirical nature. The procedure may be comparable to Whitehead's principle of "extensive abstraction" [69].

4.5.3. The complement: what we do not assert

Above we introduced the set $\mathcal{B}$ by selecting those elements of $\mathcal{A}$ which satisfy $\mho_b A$. This is enough to extract the property Ω_b once we access the class collected by satisfying $\mho_a A$. This says something about the objects exemplifying the property Ω_a, but we now ask what is the set of elements $\{X\}$ satisfying

$$\neg \mho_a X;$$

what is the negation of a property introduced by collecting elements found to pass a test devised for this purpose?

This is asking about the character of the negation of a statement (here taken to be the occurrence of a physical property). A proposition may be negated in several ways; i.e. the test for this may be constituted differently. In particular there is no set defined by writing

$$\mathcal{X} = \{X \mid \forall X; \neg \mho_a X\} .$$

The problem here is the range of applications of the quantifier $\forall$; to which universe do we have to apply the test Ω_a? Its proper use is in the formula (1), which fixes the range of entities that may exemplify a property Ω_b. The result (2) collects those elements of $\mathcal{A}$ which fail to satisfy the test for Ω_b.

Our fundamental domain for the investigation must be $\mathcal{A}$. We may always encounter many entities which fail the test for belonging to this set. We may find:

- The test is not applicable to the concept at all: "Red" cannot be applied to pressure, velocity, hopes, or love. Any formal theoretical analysis is valid only in a limited domain of total experience.
- The concept is not unambiguously definable, thus no crucial test is possible: Absolute velocity, number of clouds, and human motivations are not quantifiable in reproducible manner.
- The entity may be subjected to the test, and either the result becomes ambiguous, i.e. it is not reproducible, or it depends on the observer's own situation: Some variable may be random or like velocity depend on the observer's own situation.

After fixing $\mathcal{A}$ we may encounter a new entity Y :

$$\neg (Y \in \mathcal{A}) .$$

which definitely is found to obey

$$\neg \mho_a Y.$$

In this case we *have* learned something about the empirical reality, i.e. the complement property $\overline{\Omega_a}$ has been instantiated and the first element of a new class has been found. This may be used to extend our concepts to describe a broader range of experience. If this behavior is validated, the set $\bar{\mathcal{A}}$ may be constructed and we can extend the discussion to the class $\mathcal{A} \cup \bar{\mathcal{A}}$. Thus the physical description of empirical reality can be extended to encompass ever larger domains.

This model of concept formation hinges on the logical concept of negation. Negating the sentence:

"I have found a red object" (3)

may be either of

"I have not found a red object"
"I have found an object that is not red."

Which one holds depends on the task set as a judge of the correctness of the statement. This may be found to make (3) true, but to deny its truth, we need to know more. Thus negation of an assigned test implies an extra investigation of the intended concept. Only when this is concluded can we use both a statement and its negation. In making a statement, we need to know what its validity serves to exclude.

In formal proofs of mathematics, the conclusion proved is usually clearly stated, but it does not seem to be customary to tell what the proof claims not to be a possible state of affairs. Thus the standard diagonal proof of the cardinality of the reals only proves that all the decimal numbers cannot be numbered. To claim that this implies that their cardinality is "larger" than that of the integers implies a genuine extension of the concept of size of a set. The action is allowed as long as it may be used with impunity. We do, however, note that the existence and construction of the continuum is far from philosophically opaque [70]. The same holds even more with respect to those reals that complement the algebraic numbers.

4.5.4. Is there a total world?

Let us approach the ambitious task to describe everything there is. We proceed by catching one entity after another. To these we apply a class of tests $\{\Omega_k\}$. The entities investigated can then be assigned to sets $\mathcal{A}_k$ according to the outcomes. These sets may then be analyzed further with respect to additional tests to make each one uniquely identifiable. Let us assume that this is done by the classification

$$\mathcal{W} = \cup_{k \in C} \mathcal{A}_k.$$

When the index set C becomes large enough (infinite) we hope that the full set $\mathcal{W}$ will describe the whole world. This is taken to signify

$$\forall\Omega_k; \exists A; \mho_k A \Rightarrow \exists\overline{\Omega}; \neg\overline{\Omega}A.$$

If a physical property has been found to be possible, it must be logically possible for it not to be realized.

In addition we may require that the properties tested for by $\{\Omega_k\}$ are mutually exclusive

$$\mho_k A \Rightarrow \neg\mho_n A; (\forall n \neq k)\,. \tag{4}$$

The properties are the case or not the case independently of each other.

We now take the set

$$\mathcal{W} = \cap_{k\in C}\,\mathcal{A}_k = \{W \mid \forall k\,:\mho_k W \& \neg\mho_n W; (n \neq k)\}.$$

If this collects all information accessible about the empirical universe, it constitutes a description of all possible worlds. It includes all elements with $\mho_k W$ and excludes all elements for which $\neg\mho_n W$. When n and k independently go over all values possible we get a full description of the set $\mathcal{W}$.

The procedure fails if the description does verifiably not have a property necessary for the consistency of the world to be constructed. Remember that I am here still talking about properties as something which we know a procedure to test for. These conditions constitute the elements for a classical description of a "possible world". It remains to be seen if our empirical knowledge about reality allows us the luxury of such a description.

4.5.5. Classical approach

The set of "all possible worlds" is a complete description in terms of all possible facts. The totality of facts determines what is the case and what is not the case. Each item can be the case or not the case while everything else remains the same. These words in Wittgenstein's *Tractatus* lay the foundations for "possible world semantics". But even totally classically this program cannot be realized. An initial effort to reduce the facts to "sense data" has failed to lead to the desired goal. The facts affecting our orientation in reality do not form a denumerable sequence. We want to talk about many various entities, and it is not clear how to extract a complete set of independent "facts", and it is obvious that they cannot be accommodated into a denumerable set. When all is said and done, usually something needs to be added. This requires complementing concepts without obvious limitations. On the other hand, even if such a set is built, it is not denumerable and thus quantifying over it is dubious to say the least. This is our version of the set-theoretic statement: "There cannot be a set of all sets."

Even if all purely scientific issues have been brought into the frame of a theoretical description (which they have not), we want to discourse over hopes, beliefs, and love; all requiring totally new multitudes of concepts. There is no reason to believe that such a process will converge. The various ranges of topics will require their own nonoverlapping sets of meanings. This may be what Wittgenstein acknowledges in his concept of "language games".

4.5.6. Quantum approach

The ultimate theoretical description of scientific reality must rest on the conceptual framework of quantum theory. In this it is obscure what part is ontology and what is formalism. So far it has not been possible to derive features central to human life except within the technology of natural science. It may even be contended that it will never be possible. I may propose that what exists in science is manifestations of a universal substance called "energy". Where there is no energy above the empty vacuum, there are no processes to be observed or described by scientists. The modern energy concept emerges as the successor to the classical concept of *substance.*

An occurrence of energy offers an entity subject to scientific investigation. Such entities possess certain properties which allow us to recognize their occurrences in various contexts. Thus, e.g. the electron has a definite charge, spin, and rest mass. These variables are not appearing randomly in the empirical material, but take ranges of possible values: the spectrum of elementary particles emerges. These are the *primary properties* of possible excitations of energy.

The *accidental secondary properties* of quantum entities are given by the quantum state imposed in an experiment by the initial preparation. In quantum theory it is customary to assign the term "system" to the entities assumed given. The quantum state determines all outcomes of experiments directed at the system under investigation. The peculiar feature of the quantum description of reality is, however, that it can make only probabilistic predictions. This happens after the investigator has decided which measurement he is to perform, and intending different observational procedures he gets different probability predictions. The concept of complementarity governs which procedures are possible and which are excluded. The investigator can relate his information to a potential outcome in reality only after having decided what to measure and calculated the corresponding probabilities. Then these quantum probabilities can be related to features of reality; the reality of the associated quantum state remains shadowy at best.

The quantum description of reality is extremely well tested. It agrees with observations in all cases where tests have been possible. Where it produces numbers, these emerge with unprecedented accuracy. But it supplies a weird description of reality. All occurrences can only be interpreted as stochastic events, obeying a highly nonclassical probability calculus. The state as a property of quantum entities does not in itself allow direct observation; only its manifestations constrain possible measurement outcomes.

The quantum description of a state cannot be complete in the sense introduced in the previous section. The two-level state

$$| \Psi\rangle = \frac{1}{2}\left(\sqrt{3}\,|\uparrow\rangle + |\downarrow\rangle\right)$$

assigns the probabilities

$$P_{\uparrow} = \frac{3}{4}\ ;\ P_{\downarrow} = \frac{1}{4} \tag{5}$$

to an observable $\sigma_{\uparrow\downarrow}$. In the same situation, the state

$$| \pm\rangle = \frac{1}{\sqrt{2}}\left(|\uparrow\rangle \pm |\downarrow\rangle\right)$$

assign to its possible observed outcomes the probabilities

$$P_{+} = 0.933 : P_{-} = 0.067.. \tag{6}$$

The answer we obtain depends on the question we pose. The same state may entail different propositions in different interrogations.

Any experiment determining probabilities must have access to an assembly of identically prepared basic systems (in principle an infinite number). Measuring these we may try to verify (5) or (6). Each observation utilizes one system, but each one can only contribute to verifying one choice or the other, both cannot be verified from the same subassembly of systems. The two variables cannot be tested on the same sample.

In the situation described above, we have considered two observables (physical properties) assignable to the same system. However, we may not assign these probabilities independently. The quantum description needed to realize one set of probabilities cannot be changed without affecting the other one. Still, to give a complete description of the state of the system, the "state of affairs", we need both, and in fact even more. There is no complete description in terms of independent properties. Thus we cannot give a complete description of the actual world in terms of lists of facts.

Actually, quantum theory is even more intricate. We will have to include pairs of simple entities into the description of empirical reality. These pairs have correlates which are independent of the spatial separation of the constituents. The quantum correlations are based on the probabilistic interpretation of quantum states. These are conditioned by the amount of information imposed on the pair during the process of preparation. Thus the state we have to assign to the systems will depend on what conditions we assign to the process of preparation. If we learn more about the system here and now, we have to instantaneously adjust the expectations over the whole universe. This appears to be some queer action at a distance, but it is a natural consequence of the essential probabilistic interpretation of the formalism. This only predicts probabilities, its interpretation rests on assigning distributions to ensembles of (classical) measurements. When these are calculated, one must condition them on all information previously extracted from the system by *anybody anywhere*. The quantum probabilities are not determined by the knowledge possessed by any particular individual; they are an absolute measure of the leeway left when all possible experimental information has been used to condition the observation. In practice all this information may not be available to any observer, in which case quantum ensembles need to be introduced. They require the use of density matrices, which, however, does not affect the general argument.

Nonlocal objects can have nonlocal properties. If one constituent in a delocalized system is acted upon, the nonlocal relation may be changed. As an example I take my wife and me; we are married even if she is away in Las Vegas. If she divorces me there, I am instantaneously divorced even before the mail brought me the official papers confirming the act.

One may object that such relations exist only within the social network of human relations. It is easy to see that this is not true: The stick A in Nevada is longer than the stick B here. If someone breaks the stick A where it is, stick B instantaneously becomes the longer one. Thus even classical two-party relations may be affected nonlocally. This holds even more so if we deal with probabilistic relations. If I know the length of stick A, but I need to guess the length of stick B, it is *a priori* longer by 50% probability. When I learn here and now that it was manufactured just like A, I will know its length wherever it resides.

As a pair of entities make up a correlated quantum system, we need to represent its state by tensor products of states referring to the individual constituents. Assume we have prepared the highly correlated state of systems a and b

$$\mid \Psi\rangle_{ab} = \frac{1}{\sqrt{2}}(\mid +\rangle_a \mid +\rangle_b + \mid -\rangle_a \mid -\rangle_b);$$

such a state is called "entangled". This is a collective property if looked at from the viewpoint of the constituents; if we consider the pair as the basic quantum system, the entanglement is a property of this emerging new totality.

We test for the property A, of system a, independent of the value of B in system b, we find that they occur with equal probability

$$P_{\pm}^{(a)}((B=+)\vee(B=-)) = 0.5.$$

Thus: having no information about the observed value in b, we must assign equal likelihood to the outcomes at system a.

On the other hand, if the value of B has been measured and found to be "+", we have to assign different probabilities to the potential outcomes at a

$$P_{+}^{(a)}(B=+) = 1\ ;\ P_{-}^{(a)}(B=+) = 0\,.$$

Thus knowledge gained at b constrains what can be obtained at a even if no one there knows the outcome at b. Thus information gained imposes its outcome on further measurements; the initial preparation has acquired a new constraint.

This situation is sometimes erroneously described as a jump of the state of system a induced by the observation at b. That this is incorrect is most clearly seen from the fact that the correlation so described is independent of the actual order of the two observations. Thus no information can be transmitted through this bipartite system. This is comforting because of the threat of conflict with relativity. The relative state of motion between the two constituting systems does not influence the observed correlation.

Finally we have to inquire whether the quantum description can be interpreted as a *possible world* formalism. We have to remember that the basic unit of a system has predicates defined by the quantum state. This already excludes a full description in terms of independent properties or facts. We can measure an arbitrary component of the spin, but only when knowing the values of the spins in three directions can we assign a state to the system. However, we cannot know all these components either in preparation or observation. The outcomes of these observations cannot then be taken as independent.

It becomes worse, however. Starting from individual basic entities, we know that we can prepare quantum-correlated states which relate to the pairs as constituting complete systems. They offer many novel quantum systems which have to be assigned common properties by correlations. Affecting one of the constituents, we will significantly change the probability assignments for the other component. In addition, there is a range of possible actions, all leading to different results. The process of "teleportation" is an application of this possibility.

We find that the quantum description of empirical reality is not of the form assumed in a possible world view of reality. It is not possible to identify a basic set of facts which fully and independently fix the state of the world. The fact that the observer can select what to measure, i.e. define a "context", makes it impossible to describe even the basic units classically. In addition, we may posit combined entities which constitute states on their own right. This process seems to have no end: We can imagine a multitude of incompatible observers; there are more than two noncommuting variables. We have discussed pairs, but multiparty entanglement is possible and challenging. Here the paradoxical nonclassicality of quantum theory enters in full force.

If quantum theory describes the world of reality, this is not a "possible world". The formalism of the theory does, of course, exist and it is consistently described by the physicists. Thus it is not logically impossible, but interpreted as the existing world it does become paradoxical. In its own realm it is perfectly consistent, but its message as picturing the structure of all reality is highly muddled.

4.6. Physical applications

4.6.1. Numbers of objects

The simplest domain of mathematical reasoning appears to be the integers. They start from the concept of zero and the procedure of going from any given number to its successor. This defines the operation of addition. The ensuing sequence is without end, but it is an arbitrary act to call it infinite. No statement about all integers can fundamentally mean anything but the fact that the domain is not exhausted by any finite set of numbers. As an object of mathematical interest, the integers form a world of tantalizing complexity and surprising beauty. The best-known enigmas concern the fully determined but apparently random prime numbers. One may ask: How can the simple process of "adding one" create all these structures?

As empiricists we may ask how it is possible that the concept of number appears so natural? Here it seems that the emergence of such a concept rests on one of the simplest persistent facts of nature. There are assemblies of physical object similar enough to allow them to be identified, gathered into sets, and reidentified over time and by different observers. We have heaps of five stones, which can be mapped reversibly on to heaps of five sticks, and so on. This certainty that there are physical objects is a pure experience and needs no deep justification. The mapping between heaps, on the other hand, is a basic operation which allows us to judge which finite sets have the same number of elements. We may thus introduce an equivalence class and call this a numeral, "five" say. Thus we have given a name to the "cardinality" of any finite set that we may be able to identify with certainty. Ephemeral objects like clouds cannot objectively be counted and thus their sets cannot be assigned a number.

In number theory, we usually use the recursive method: Something holds for $x = 1$ and from the validity of the result for y follows that it holds for $(y + 1)$. Then we seem to conclude that it "holds for all x''. The example below is a proof that within the standard axiomatization

$$a + b = b + a. \tag{7}$$

Proof. The associative law is supposed to be postulated:

- $a + 1 = 1 + a$, obviously hold for $a = 1$.

  If it holds for n that

  $$n + 1 = 1 + n$$

it holds for $n + 1$, because

$$(n + 1) + 1 = n + 1 + 1 = 1 + n + 1 = 1 + (n + 1)$$

- Repeating this procedure we arrive at $a + b = b + a$ as required.
- From this we presume to be justified in claiming that it holds for "all numbers".

What the "proof" shows is, however, only that we have a procedure to demonstrate the validity for any particular pair a, b. To claim the generality of the outcome is only to state the obvious fact that our procedure for constructing the integers has produced this result for any particular case. We have a result following only from the consistency of our construction of the integers and their algebraic manipulations. The illusory character of the induction has been pointed out by the mathematician Poincaré [70].

Wittgenstein spends much effort on discussing proofs of mathematical theorems. Unfortunately he chooses as prototype expressions of the form $2+3=5$. This is, however, not a theorem at all; there are no autonomous entities 2 and 3 which combine to emerge as the novel entity 5. But we meet another example of the consistency of our procedure defining the integers. The "verification" of the result is namely

$$2+3=1+1+3=1+4=5.$$

This is simply based on the process of adding one and identifying the new integer at each step. A mathematical theorem must be universal, i.e. it must hold over some domain as does the result (7). But then we are back at the problem of quantifying over infinite sets. Induction only shows that we may draw a conclusion for *any* particular case; to restate the result for **all** cases is either a major ontological commitment or a trivial rephrasing of the results proved.

The arguments above may seem trivial, but they reveal deep relations underlying the introduction of the integers as existing entities. Finite results are often obvious, but universal statements for all numbers are intriguingly difficult to prove. Examples are offered by "Fermat's theorem" and "Goldbach's conjecture". But the sets of (potentially infinite) numbers of physical objects induce us to introduce integers and their algebraic properties. This is a perspicuous example of how a robustly appearing feature of empirical reality allows the formal construct of a framework which consistently can be assigned real existence and hold to support highly nontrivial properties within its realm of accepted entities and their manipulations.

4.6.2. Space

We feel that our life is taking place in a space with three degrees of freedom: We may move forward, deviate to left or right, or jump up or down. Our possible movements are guided and restricted by the spatial structure of our environment and the objects we encounter there. The very concept of a classical object implies that it occupies a volume of space and displays a durable spatial structure. We feel the features of the materials surrounding us and our visual pattern recognition is trained to coordinate the visual environment with the tactile experience of our bodies. We live our lives in a three-dimensional manifold.

Logically this is not obvious; a world of different dimensions would have to be described by a different geometry, but it would still fit into our conception of a possible reality. There are no compelling arguments why the dimensions are three and only three. Most physical theories can be separated in independent phenomena along the different axes of space. Only phenomena describing whirls, e.g. vortices and electromagnetism, depend essentially on the fact that screws need just three dimensions.

But even a single dimension of space offers its enigmatic challenges. Mathematically one may claim that the direction described by an axis consists of a sequence of points lying straight, one after the other. But here is the crux of the naive view: In order to introduce a numerical measure along the distance we need to devise ways to attach outcomes of measurements to the distance between points marked on the line. The procedures of mathematics construct the rational numbers by repeatedly setting off unit meter sticks along the line. But this gives only the rational points, which can be shown always to be separated by innumerable irrational points. Thus mathematics claims that there are more points on the line than can be reached by any countable procedure. This is the problem of the continuum. Classically Euclid never regarded *lines* as consisting of individual *points*; these concepts are essentially different primitive elements in his theory. Only well-defined points would play a role. Two lines meet in a point, a line cuts a circle at most at two points, and so on. These are emerging facts, not axiomatically induced consequences.

Modern set theory wants to consider the line as a collection of all its points. Then again we encounter the fact that proving something for an arbitrary point on the line does not necessarily allow us to state that the result holds for all points on the line. A simple example from planar geometry may illustrate the dilemma:

Pick an arbitrary point on a line. Construct the line going through this point being orthogonal to the original line. This defines a new continuum set for any starting point. Does this allow us to claim this to hold for all points on the line? It may seem so, but consider the union of all sets constructed from all the points on the original line. This covers the whole plane, so running along the original line we constitute the two-dimensional plane. Looking at this point by point, it is identical with sets defined in infinitely many ways. It is essential to remember that different methods to construct the two-dimensional plane do not *a priori* generate the same plane; to identify them is an intuitive act by the mathematician. But having access to this set, we cannot recover the procedure of construction or even find the line originally given. Not only is the plane reachable in infinitely many ways, it does form the basic structure where the whole geometric argument plays out its drama. Trivial as it is, this example shows how confusing it may be to quantify over a set having the cardinality of the continuum.

The introduction of the mathematical continuum has been criticized by Poincaré [70] and Weyl [71] and especially [72]; Brouwer flatly denies us the use of points we do not tell him how to find and identify [73]. But this is the case for the overwhelming majority of points on a line. And still we want to be able to tell the distance between two points with arbitrary accuracy, with an arbitrary number of decimal digits. How is this possible?

Different methods are used to justify the irrational points complementing the rationals. If we have two different methods, Dedekind cuts and Cauchy sequences say, do we introduce two different continua? This conclusion is avoided by utilizing the properties of the results obtained; they are structu-ally isomorphic, what is true in one is true in the other. Thus by identifying them, we can never encounter any inconsistencies. The construction becomes unique.

All measurements consist of comparisons between an investigated object and a gauge reference. This is described by Bridgeman in his *operational* approach [74]:

> To find the length of an object, we have to perform certain physical operations. The concept of length is therefore fixed when the operations by which length is measured are fixed: that is, the concept of length involves as much as and nothing more than the set of operations by which length is determined. In general, we mean by any concept nothing more than a set of operations; *the concept is synonymous with the corresponding set of operations.*

This defines the meaning of the concept of "length" as identical with the operation which determines its numerical value. Physically speaking, the distances are determined by meter rods not only of given length but with subdivisions given by marks along their length. Trying to find a measure for the separation of two points somehow marked on a given line, we put the measuring rod along the line and read the number corresponding to the difference between the marks at the one end and the mark of the separation to be measured. Thus we transfer the mathematical problem of the continuum to the physical problem of the marking on the rod. This seems to be no progress; we have replaced the unknown distance by an imprecise meter reading on the rod. Where is the advantage? We must consider the fact that physical objects can be moved around at our will. We can take the rod to another location and read off the length of another separation by the *same measure.* Of course the reading is highly inaccurate on an absolute scale. But this inaccuracy is the same over the whole realm of our investigation, and all distances are put on an equal epistemological foundation. Improving technology, we may construct more and more sophisticated devices for the measurements, and all improvements will be seen to follow the same pattern: When different observers measure the same distances by their various methods, once the unit of length is chosen, the readings agree and the physical distances are assigned numbers with an ever-improving technical accuracy. Distance is found to offer a durable physical concept to be taken over by theoretical physicists. This does not solve the mathematicians' problem of the continuum, but shows how practical science overcomes the difficulty.

But there are surprises along the path to ever-improved measurement accuracy. Einstein's theory of relativity shows that the invariance of the velocity of light merges the assignment of numbers to distances with the observations of time. When fundamental measurement procedures are considered, space and time become dependent in a way not foreseen by the scientists. Different observers have to use different numbers to tell the distance between separated events, but the values are determined by strict rational rules so no confusion needs to emerge. The assigning of numbers to time intervals is discussed in the next section.

But the theory of relativity assumes the existence of perfectly rigid rods to be used as meter tools. To posit the existence of such objects is clearly seen to be inconsistent within the theory they serve as a basis for: If we have a perfectly rigid rod of length L, we can start pushing one end and thus force the other end to start to move instantly. This transmits the information of our push over the distance L in no time, which clearly conflicts with the theory we are supposed to build up. Thus the ideal rod used to define the theory cannot exist.

When quantum mechanics is added, the problem becomes more acute: A solid rod is presumably built from atoms joined by chemical bonds. These act as springs having their zero point energy different from zero. The corresponding uncertainty in the length of the spring makes each step from one atom to the next not totally defined. Adding atomic separations to form a macroscopic rod, we easily find that the expected distance between the marks on the rod fails to satisfy the requirement of providing a robust measure which remains the same independent of how and when it is used as a measuring scale. Quantum theory does not allow the existence of rigid solid rods.

In physics this dilemma is solved in an ingenious but exotic way. When a liquid becomes solid, its structure is changed by a phase transition. This introduces an order parameter describing the geometric structure of the ensuing crystal in an ideal way. Each atom is assigned a well-defined lattice point following from the geometry of the long-range order imposed by the order parameter. The atom seems to take cognizance of the place where it ought to be, and it may only experience its quantum uncertainty as relating to this position. As a result (nearly) all atoms build up the lattice corresponding to the symmetry of the ordered state. This order is seen when the crystal is investigated by X-ray scattering. Atomic fluctuations affect the result only by a minor blurring of the scattered data, but the geometric pattern of the emerging radiation gives decisive information about the lattice structure. This structure, in its turn, tells how big the solid body is, and thereby allows the introduction of measuring rods of definite repeatedly used lengths. Thus, via the theory of phase transitions, quantum physics allows the distances to be measured as in the classical situation. The ideal concept of a durable measuring rod can only be supported by reference to subtle physical considerations. But these are never exact and only within the fuzzy but well-controlled procedure of measurements can we assume the measure of space to be realizable.

Our empirical procedure does not allow us to state anything concerning the ultimate structure of spatial points. Thus some of the more exotic theoretical speculations may one day provide new insight into what it means to be an individual point in the three-dimensional universe.

4.6.3. Time

In modern physics, time is often considered as an additional dimension of reality, in many ways having the same status as the spatial measures of space. This is motivated by Einstein's theory of relativity, which shows that measurements of time and space intervals become entangled. The finite velocity of light prevents any universal assignment of numbers to the intervals but presents the result as a totality experienced differently for each observer. However, the theory determines relations between the results obtained and restores rationality to the recorded experience. Superficially the time and space intervals form a four-dimensional manifold where the standard rotational operation is replaced by the seemingly analogous Lorentz transformation.

This presentation is, however, an illusion. Different observers may not agree on the coordinate identified as the time, but each one can uniquely single out a time direction in his world. Time flows, we live in the present and past, and future can exist in planning and imagination only. We do not even experience the flow directly, but only realize that what has passed is gone irretrievably. In this aspect, space differs radically, we can go back to a location we visited previously, but we cannot visit the past nor the future, however attractive this appears in fiction.

But the role of time is more profound still. Our best physical theories provide us with causal inferences making possible our interpretation of the past and precautions for the future. The theory does allow causal interpretations only if events are temporally ordered as seen by all observers. Thus the events emerging order the occurrences into causal sequences which for each observer uniquely defines a progression of time in reality. The observers will assign different measured numbers between the events but their order will be universally agreed upon. In particular, no action can turn time backwards nor change the direction of time between causally related events. Living in the frame of his own world, each observer organizes his experience into a consistent and rational ambience where in the limit of small relative velocities, the classical description of reality emerges.

But the experience of time is a subjective feature of our lives. Different observers need different time concepts, and our consciousness is notoriously unable to make accurate determinations of time elapsed. Waiting for a bus, we experience minutes as an eternity whereas an evening with merry friends passes but all too fast. How is it that we can still posit a concept of time and assign a definite measure to its flow? In order to find an answer to this, we have to look into the demands we set on phenomena if they are going to have the status of a scientific entity.

Time forms an integral part in the history of science. The heavenly bodies offered an early demonstration of change and recurrence. By observing the moon and the stars, early man could regulate his life according to his expectations of rains and floods. Early it was recognized that natural phenomena were reliable measures of the progress of time. The roots of science and mathematics are found in these early astronomical and geometric investigations.

Having grasped the concept of time, human inventors developed artifacts to measure it. Flow of sand or water was used to delimit slices of time. Also the periodic vibrations in nature suggested the use of the pendulum for the same purpose. From here developed the concept of a clock mechanism, which still survives in many clocks. Thus simple physical systems have been shown to display regular repetitions of their earlier configurations.

Even after the appearance of these different time-measuring devices, how can we posit the existence of a real physical time entity, which proceeds uniformly forwards, and whose measure is supplied by our measuring clocks? As time proceeds in one dimension, it makes no sense to attempt the superposition of two different time intervals in order to compare them. There are no meter sticks for time. We cannot slide one standard interval over another to affix the measure with regard to a definite standard. In the Euclidean version of space geometry this operation serves an essential role, which cannot be taken over into the geometry of time. Our clocks all tick their way into the future, and we can have no absolute standard to put against an arbitrary slice of time.

We can, however, still introduce a standard unit for time. This derives from the observation that time can be counted by a multitude of physical recurrences occurring over shorter or longer intervals. These are used to construct clocks with technologically more and more sophistication. A duration identified by the separation between observable occurrences can be measured by different clocks, based on different physical phenomena. We will then find that the time separation between event A and event B compared with the separation between event C and event D gives the same ratio independently of how the time has been measured physically. The periods of the moon contain the same number of days independently of whether time was measured by the flow of sand or the swings of the pendulum. It becomes a contingent fact that all periodic phenomena progress in a synchronized manner.

Time itself takes no absolute values, but it was observed to flow with the same rate whichever physical process is used to assign a number to it. There is no *a priori* measure of the duration of time, but physics allows the introduction of a uniform continuum of instants. Any absolute measure of the flow is not assignable but the uniformity allows the imposing of a universal reference, arbitrary but stable.

Hence we find that the introduction of the concept of physical time rests on the stability of some feature of empirical experience. Like number and separation, time measures can be introduced because there is a systematic agreement about its flow rate at different times and by different observers. This is made complicated by the theory of relativity, but the pattern is not broken; measuring time by different observers gives results related uniquely by the theory.

Going to general relativity, the phenomena become more incomprehensible. However, the human observer lives in a nearly Euclidean space, and then the classical point of view can be retained. We cannot imagine what time concept would emerge to a rational observer living at the edge of a black hole. But this is presumably for ever to remain a meaningless question.

Time in quantum theory is, on the other hand, complicated in a different manner. Time is not an observable like in Newtonian physics; it just runs its course and the theory assigns to it the probabilistic causality which governs the predictions of future experiments. So far there is no mystery attached to the concept. The enigma emerges when we contemplate how an observation here can instantaneously condition measurements performed very far away. No concept of causality is allowed to do this. The answer lies in the statistical nature of quantum predictions; if we learn more here and now, we instantly need to readjust our predictions for future outcomes everywhere in the universe. Thus the character of quantum time is intimately connected to the intriguing question of how the formalism of quantum theory is interpreted. And this is definitely a different story.

4.7. What the theoretical structures tell us

4.7.1. Quantum theory revisited

The theoretical models are not pictures of reality (mirrors of nature) but demonstrations that reliable maps of parts of reality are possible. Relations between theoretically introduced concepts define what we are able to say. This involves the algebraic structures. What can be taken to exist depends on the possible representations these algebras have. This allows us to introduce the properties of elementary particles.

We want to describe some phenomenon in the microworld. We use our theoretical concepts to achieve this. But the concepts used have to be founded in some process giving them meaning. However these, in turn, must be defined by other concepts. This goes on until we encounter concepts that can no longer be founded; they have to be understood intuitively. In physics, these are the concepts of classical physics. Hence Bohr insisted on descriptions in terms of classical explanations. The classical limit of quantum theory is statistical mechanics, i.e. the analysis *à la* Bohr must always be carried out on the limiting classical description.

The Russian authors state [75]:

> We can assert that the statistical character lies in the nature of things. But we do not know the deeper reasons lying at its foundation, or more precisely, we cannot say what exactly is the relationship between the statistical character and other deep natural phenomena.

Measuring consists of two acts: First a real interaction changing the way things are. Secondly out of the ensemble emerges a certain definite quantity which is, in fact, reified. This does not tell how things are, but only changes our knowledge of observable relationships.

Physics forms concepts by devising tests. Object a belongs to set S_a if it satisfies the test $\phi(a)$. Thus all sets are finite (or at most denumerable). This does not mean that we know what the complement set is. We still have to learn when the concept is not applicable, $\neg\phi(a)$. This only allows us to create world descriptions.

In quantum mechanics, some physical concepts are considered unreal because they cannot unambiguously be assigned numbers. "Momentum does not exist because it takes no definite value." But this does not invalidate the concept as such. In the proper experimental setting, it still plays a role; we may still have validity of conservation laws.

If the state vector is a record of our knowledge, how do the individual objects know how to act so that the final ensemble averages predicted are realized? This is the ultimate question for understanding the probabilistic character of quantum theory.

The knowledge we talk about is not "somebody's" knowledge but accessible knowledge. It is verifiable by anybody who cares to look. Here time comes in, the test of the information available may require time to emerge as "classical" information. Only then can the possible correlations be established. But classical information can only be transferred according to relativity's limitations. The requirements of relativity belong to the classical realm and thus become paradoxical when transferred to the quantum domain. *Do not apply classical concepts in the quantum realm!* Classically they appear contradictory; in quantum theory they would be complementary.

4.7.2. Truth and consequences

The concept "truth" is not easily applicable to anything. Facts of our human ambience are manifest, they just "show themselves". They are not true as such. This was discussed in Sec 1.2.4 above.

Within formal systems, proofs may be taken to supply true statements. This view is very limited as Gödel has proved. However, within a symbolic representation of some feature of reality, we may find enough empirical evidence to assert certain hypotheses as justified. This is "warranted assertibility" only. No statement pertaining to the formal description of empirical evidence can be assigned as "true" in an absolute sense. The concept "truth" is employed rationally in the following cases:

- Direct observations, sense data generalized to (numerical) outputs from experiments, are not "true" or "false". They are manifest, i.e. they cannot be meaningfully denied; they allow no counterfactuals. It is, of course, true that "they are manifest", but that elaboration adds nothing to the discussion. It is but part of the background to living and acting.
- In formal mathematical systems, some things are "provable" consequently *true*. By extending the framework, we may even be able to prove some Gödel sentences. But then they are true in an extended system. We know that this process does not terminate. Thus there is no limiting system that provides Platonic truth in formal systems.
- If we have a formal structure being a model of some physical realm of phenomena, all sentences cannot be logically decidable. There have to be some which are *contingent*. Gödel's incompleteness is essential if we want to

describe nature by formal models. This is necessary if the theory is to apply to an empirical activity. Some statements of the theory must be contingent and thus falsifiable. If they are tested they may be called "true" in the formalism. A better word is *warranted assertibility*. Because the frame may be changing or extending its realm, these judgements are not absolute.

- All knowledge acquired by mankind concerns limited realms of experience. We do not have a genuine "theory of everything". We do not even believe that such an all-encompassing framework is possible; you cannot make a true map of the earth on a two-dimensional sheet. To quote Weyl [68],

> ... the propositions and laws of physics, taken individually do not have a content which can be verified individually; it is only the theoretical system as whole which can be confronted with experience. What is achieved is not intuitive cognition of an individual or general state of facts, and a description which faithfully portrays the given conditions, but theoretical purely symbolic construction of the world.

The openness of intuition guarantees the possibility to amend our theories when novel experience is encountered.

It is of sufficient importance that physics has always admitted a loophole in the necessity of Nature.

5

WITTGENSTEIN ENTERS THE SCENE

5.1. Life and work

Ludwig Wittgenstein [76] was born in Vienna on April 26, 1889 as the fifth and youngest son; he also had three elder sisters. His father was a wealthy industrialist and his mother brought great artistic talents to the children. The family was of Jewish origin but they had adopted Christianity, and religion played no strong role in their life.

Wittgenstein went to school in Linz, and in 1906 he moved to study technology at Berlin-Charlottenburg.

In 1908 Wittgenstein moved to England, He spent the summer in atmospheric research in Derbyshire and enroled as a student of engineering at the University of Manchester in the autumn. Here he remained until 1911, and he worked on the design of propellers and the idea of a jet-driven airplane. As part of his engineering training, Wittgenstein became ever more interested in mathematics, and he started to penetrate its foundations by reading the works of Bertrand Russell and Gottlob Frege. After a visit to Frege, he decided to study with Russell, and entered Trinity College in Cambridge as a student.

Wittgenstein's father died in 1912, and he inherited a great fortune. In accordance with his ascetic view of life, Wittgenstein gave away his assets, which later made his economic situation precarious.

In 1914–18 Wittgenstein served in the Austrian army until he became a prisoner of war in Italy. At this time, he already had accumulated parts of the material to become the published work *Tractatus Logico-Philosophicus* [10]. During the war, in imprisonment, he worked further on this and finished the manuscript.

In 1918 Wittgenstein managed to send a copy of his manuscript to Russell, who wrote a foreword which did not please Wittgenstein. He met Russell in Holland to discuss the publication of the manuscript in 1919. This was finally published in German in the journal *Annalen der Naturphilosophie.* It was published in English with the parallel German original in the following year. This was to be the only Wittgenstein book published during his lifetime.

Wittgenstein considered his published work to put an end to philosophy, and consequently he turned to other activities. He trained as an elementary-school teacher and worked as such in small villages in lower Austria. He was not a success at working with young children, mainly presumably because of his lack of patience with intellects slower than his own. Consequently he also worked as a gardener and contemplated entering a monastery.

In 1926–8 Wittgenstein worked as an architect, designing a house for his sister in Vienna. This was an original creation in the functionalistic style propagated by the architect Adolf Loos, whom Wittgenstein knew and admired.

In 1928 Wittgenstein listened to a lecture by L E J. Brouwer, the main father of the intuitionistic school of mathematics. Its title was "Mathematik, Wissenschaft und Sprache". It has been claimed that this induced Wittgenstein to take up philosophical research again.

Wittgenstein returned to Cambridge in 1929, where he registered as a student. However, his book and the earlier stay at Cambridge enabled him to achieve a doctorate in the same year. He was then made a Fellow of Trinity College.

In 1935 Wittgenstein visited the Soviet Union and planned to settle there, but in 1936 he moved to Norway and began the revision and organization of his thought; after his death this material was to be published as *Philosophical Investigations* (in the year 1953) [77]. Wittgenstein was elected the successor of G. E. Moore who retired in 1938.

War broke out in 1939. Wittgenstein was, by now a British citizen and worked in hospitals in London and Newcastle.

In 1947 Wittgenstein left his position as a professor and the position was eventually filled by G. H. von Wright.

Wittgenstein moved to Ireland, quite separated from most human company. After falling ill, he remained with friends in Cambridge and Oxford. He wrote profusely during this period, but no material was prepared for publication. Wittgenstein's notes started to appear under the editorship of Rush Rees, Georg Henrik von Wright, Elizabeth Anscombe, and others.

Wittgenstein died of cancer on April 29, 1951 in Cambridge. His last words are said to have been: "I have lived a happy life."

I make no attempt to document the publication history of Wittgenstein's notes and the chronological development of his thinking. I only list the published collections which have been available to me during the writing of this text [77]–[92]. The fact that he published nothing of his later work himself, combined with his rather unsystematic oral presentations, leave many issues as objects for active research. Many accounts of the continuity and developments of his philosophy are to be found in the profuse published literature commenting on aspects of Wittgenstein's work.

5.1.1. Background for Wittgenstein

Like so many obscure features in the life of Wittgenstein, the philosophic influences on him remain unknown. He was, no doubt, familiar with the ideas of Kant, who dominated the thinking of the whole intellectual atmosphere of the German-speaking cultural environment. Remarkably enough, when Wittgenstein in 1931 listed the writers that had influenced him, he did not mention Kant. On the contrary, he put the physicists Boltzmann and Hertz first, and philosophy proper was represented by Schopenhauer.

In Vienna, a multitude of questions were topical when the nineteenth century turned into the twentieth. Among these, the issue of scientific progress became important for the physics of the new century. The dominating figure in the philosophical discussions in Vienna was Ernst Mach (1838–1916). He regarded himself a physicist, but his main influence was of an epistemological character.

Mach followed the tradition of the British empiricists Locke and Hume. He claimed that the world of humans must be constructed from experience; he used the term "sensations", but took this term in a broad sense. The only things available were these data. Out of them scientists had constructed metaphysical entities beyond what was justified by the observations. His aim was to eliminate all such constructions. For this he applied a principle of economy. The theories of science are summaries of experienced facts, their form determined by the organization of mind. Conceptual economy would provide the simplest mental images summarizing the empirical material. That is to "save the phenomena".

Mach presented his philosophical outlook in the Preface of his work *Die Mechanik* (1883). Here he sets up a set of theses, which were to influence generations of thinkers of various ilks:

- The simple facts to which we reduce the more complex ones are in themselves always incomprehensible, i.e. not further analyzable.
- No basic fact is more understandable than another. The choice of basic facts is a matter of comfort, of history, and of habit.
- Bodies are nothing but bundles of reactions connected by law.

These statements foreshadow the later approach by Wittgenstein. The ideas also came to have a strong impact on the young Einstein and later on the creators of quantum theory.

Mach's ideas were developed in competition with Heinrich Hertz (1857–94). He was more of a Kantian than Mach. In order to solve the problem of the emergence of the world, he followed Kant in positing frames ("categories") supplied by humans to sort the experience into systems, theories. He also published a book on mechanics, *Die Prinzipien der Mechanik in neuem Zusammenhang dargestellt* (1894). This book appears to have reached a broader physics community than Mach's work. It also had a decisive influence on Wittgenstein's attitude to scientific theories.

Hertz's mechanics was based on what he called "pictures" (*Bilder*), which would, perhaps, be called models today. His models had to be based on three principles: They had to be internally consistent, i.e. not conflicting with our logical ability to envisage a theory. They secondly had to be correct, which means that they must agree with the observational facts. There might, however, exist different models which satisfy these two criteria. Then the principle of economy might be applied: the model having the smallest number of arbitrary presumptions was to be preferred. Hertz realized that no theory could be totally without such elements.

Comparing Hertz's outlook with that of Mach, the former provided a more adequate framework for the development of physics. Mach's elimination of concepts and theory constructs was entirely based on the analysis of the human mind. It introduced a *psychophysical* framework, which assigned the drama of physics to a mental scene. No criticism of the theory as physics becomes possible thus. Hertz, on the other hand, allowed an investigation of the theory from within. Thus the limits or restrictions of the theory could be discussed in a fruitful manner. Hertz reached this position from his attempts to understand the then novel theory of Maxwell. The electromagnetic phenomena offered a rational extension of the mechanical formalism into a new direction which needed to be understood.

Ludwig Boltzmann (1844–1906) worked more deeply on scientific issues than Mach. He was to be one of the founders of modern statistical physics. In addition to his physical works, he presented his thoughts on science in essays and talks. These were published as *Populäre Schriften* in 1905. Boltzmann was recognized as a thinker merited enough to be appointed Mach's successor in 1903.

Boltzmann was more of a realist than Mach. This does not mean that he did not appreciate the necessity for observational terms or the antropomorphic status of even the best of our theories. The attitude of a physicist to matter is always schizophrenic. Accepting that only experiments can teach us about facts, he has to assume that these are facts about something; the space and time must persist and matter must somehow reside in them. In particular Boltzmann differed from Mach in assuming that atoms exist as elements of reality. He states:

> It is just here that we are in an area where a whole new kind of methodological speculations has grown up which were given the most precise and ingenious expression by Mach, who states straight out that the aim of all science is only economy of effort. With almost equal justice one might declare that since in business the greatest economy is desirable, the latter is simply the aim of shops and money, which in a sense would be true. However, when the distances, motions, size and physico-chemical properties of the fixed stars are investigated, or when microscopes are invented and with their help the causes of diseases are discovered, we would hardly wish to call this mere economy.

Thus the stage was set both for Wittgenstein's philosophical outlook and the spectacular developments of twentieth-century physics.

Of established philosophers, Wittgenstein mentions only Schopenhauer. He had, partly in polemics against Hegel's holistic system of the world, developed a view of reality as our image only. However, the universe and its life were propagated by the inner forces he termed "Will". In his book *Die Welt as Wille und Vorstellung* (1819), Schopenhauer presents his world view in beautifully formed prose consisting of aphoristic sequences of statements. There is no doubt that this manner of presentation appealed to Wittgenstein, who was to become a master of this art. Schopenhauer also presented pessimistic and misanthropic views of life. These also suited Wittgenstein in his most depressed moods. Systematic theories were not among the strong points of Schopenhauer.

A rather strange influence is, according to Wittgenstein in 1931, the odd personality of Otto Weininger, who published his controversial work *Geschlecht und Charakter* in 1903. Weininger was a Jew and a homosexual, who attributed all ills in society to the influence of women. They were unable to recognize truth and evaluate the consequences of actions. Hence they were totally dependent on men to manage in the world, but at the same time they asserted a totally destructive influence on their surroundings. Sexual desires were uninhibited and prostitution a self-evident conclusion. In order to achieve something the ingenious man must overcome his sexual lust. To overcome this, the ultimate solution might be suicide. Weininger himself chose this manner of exit.

It is difficult to estimate why Wittgenstein let himself be influenced by Weininger. The simple solution that it provided Wittgenstein's attitude to sex cannot be the whole truth. He said later that he took the book with a negation sign in front of Weininger's opinions. This may well indicate the truth. Weininger's serious attitude to life and his will to draw the consequences of moral dilemmas corresponded to Wittgenstein's own. It may well be that Wittgenstein did not learn the views from Weininger but rather his manner of discourse. Weininger's influence on Wittgenstein will be analyzed further below. Be that as it may, without Wittgenstein, Weininger might well be totally forgotten today.

Wittgenstein was also influenced by the philosopher-psychologist Willam James (1842–1910). In 1912 he read James's book *Varieties of Religious Experience* (1902), where Wittgenstein recognized the longing for unity beyond our technical machinery of description. James states:

> The most obvious characteristic, whereby I label a mental state as mystical, is negative. Those who have personal experience of it immediately say, that it defies every expression, that no conclusive description of its contents can be given in words. From here follows, that its nature must be directly experienced; it cannot be communicated or transmitted to others.

Later Wittgenstein got acquainted with James's *Principles of Psychology* (1890), where the possibility of introspective analysis is presented. This was early supposed to provide the only method to investigate the dynamics of the psyche. It contrasted strongly against the philosophical tenet that we only have access to sense data, psychic atomic sentences. James regards the process of the mind to be representable as a stream of consciousness. It represents a totality, whose parts do not necessarily carry meaning on their own. This concept has been claimed to influence many authors of the early half of the twentieth century. James Joyce's *Ulysses* (1922) is supposed to offer one extreme example of this genre.

Wittgenstein used James's book for examples of misuse of language in his lectures on Psychology. It is not clear exactly when he became acquainted with this work, but he certainly found it enough of a challenge to use as a basis for his discussions.

Wittgenstein had the professional training of an airplane engineer. He thus knew the principles of mechanical engineering well. There are, however, no indications that he had looked into the more modern branches of physics. G. H. von Wright confirms that Wittgenstein never took up such topics. This agrees with his general attitude that science is all right as it is, and no philosophical argument needs to or, indeed, should interfere.

It is still, however, strange that Wittgenstein does not even occasionally refer to the electrodynamic theory of Maxwell. This was well established, its utility was well asserted, and it forms an interesting abstract representation of an important area of human discourse. Even less satisfactory is his total neglect of all of modern physics. He shared his sojourn in Cambridge with the other genius Paul Dirac, who had the most astonishing intuitive insights into the structures of the modern theories. One can but imagine the possibilities of listening in on a discussion of theirs. Unfortunately, all we know about them indicates that they could not even have started to communicate.

As a student of technology in Manchester, Wittgenstein got to use advanced mathematics. During this work he got acquainted with Bertrand Russell's *Principles of Mathematics* (1903) and started to ponder over the basics on which mathematics was founded. Professor Gottlob Frege (1848–1925) in Jena had already in 1874 started to publish his pioneering work on the logical foundations of mathematics. Russell was to find serious problems in Frege's program, which inspired him to his extended effort together with Alfred North Whitehead to found mathematics on logic.

Wittgenstein visited Frege in Jena and was advised to start to work in Cambridge with Russell, whom Frege regarded as his most promising successor in the field. This advice led Wittgenstein to abandon his career as a research student in Manchester, and at the beginning of the fall term in 1911, he appeared in Cambridge and requested a position as collaborator with Russell. The details of their dealings are richly documented in the literature and can be omitted here. Suffice it to say that Wittgenstein was no humble student or pleasant collaborator. One may surmise that it took all the tolerance of Cambridge and the integrity of Russell to accommodate the antics of Wittgenstein. Up to the First World War the collaboration continued, and after the war, Wittgenstein must have found Cambridge inspiring and tolerable, because he eventually returned there.

Before that, however, Wittgenstein fought in the First World War, became a prisoner of war in Italy, and wrote his major work *Tractatus Logico-Philosophicus*. This was understood by few, and he found it difficult to be published, but eventually the work earned its author a doctorate in Cambridge. After that Wittgenstein never published anything. The work appearing as *Philosophical Investigations* (1953) was (almost) prepared for publication when he died [77]. He left a *Nachlass* of numerous notes and manuscripts which have become published over the years in more or less systematically edited collections.

There is thus ample material for the study of Wittgenstein's philosophy. Consequently a multitude of books have appeared, and it is impossible to obtain a coherent picture of his work; maybe he did not really strive to achieve this. He certainly was convinced that he never reached this himself.

5.2. Influences on Wittgenstein

5.2.1. Reading in philosophy

It seems clear that Wittgenstein had done no systematic reading of philosophy. This, however, does not imply that he lacked knowledge about its teachings. He certainly knew the tenets of Kant; that was inevitable in his circles. As a young man he also read Schopenhauer. The scientists of the time in Vienna were oriented towards fundamental issues. Thus Wittgenstein got familiar with the writings of Hertz and Boltzmann; in fact he had wished to study under the latter, but Boltzmann's death prevented this.

Wittgenstein appreciated William James much, but his initial contact was through the book *The Variety of Religious Experience.* Only later did he read the *Psychology.* The religious approach was natural, because Wittgenstein held an unceasing fascination for confessional autobiographic writings. Thus he is known to have read St Augustine's *Confessions*, Kierkegaard's works, Dostoevsky's *Notes from the Underground*, and Tolstoy's *Gospel in Brief.* The last work became important to him when he in 1916 was contemplating his own life fighting at the front. He was eventually getting impatient with the excessive outpouring of words by Kierkegaard: "He is so long winded; he keeps saying the same thing over and over again."

In the year 1931 Wittgenstein stood at the threshold of his new approach to philosophy; he listed the persons who had influenced him: Boltzmann, Hertz, Schopenhauer, Frege, Russell, Kraus, Loos, Weininger, Spengler, and Sraffa. The list is supposed to be in chronological order; they had affected Wittgenstein in this succession.

The first three names are obvious, Wittgenstein got acquainted with Boltzmann, Hertz, and Schopenhauer already at high school. Frege and Russell influenced his thinking starting from about 1911 well into the 1920s. Kraus and Loos catalyzed the Viennese cultural atmosphere for the young philosopher. Spengler's work appeared only in 1928–31, so his influence is late. Sraffa was a friend and discussion partner from the Cambridge days.

The only remaining obscure name is Weininger. He had a strong position in the discussion during the pre-war years in Vienna. This reputation was not indisputably good. Later he was labelled a bad philosopher and a propagator of prejudice and vulgarity. However, in some way, Wittgenstein found him interesting and challenging. He recommended Weininger to G. E. Moore by saying:

It is not necessary or rather not possible to agree with him but his greatness lies in that with which we disagree. It is his enormous mistake which is great. i.e. roughly speaking if you just add "$\sim$" to the whole book, it says an important truth.

The symbol "$\sim$" is Wittgenstein's sign of logical negation. So he does not negate any particular statements of Weininger's but the whole text as a single proposition. We want to try to learn how to understand this statement. In order to do so we need to take a closer look at the views propounded by Weininger.

5.2.2. The role of Weininger

Life and work of Weininger

Otto Weininger was born in 1880. In the spring of 1903, he published his book *Sex and Character* (*Geschlecht und Charakter: Eine prinzipielle Untersuchung*). Its immediate impact was only modest, but all this was to change. In October the same year, Weininger shot himself in the house where Beethoven had died. Thus he paid tribute to the greatest embodiment of his conception of a genius. This symbolic act made Weininger a legend in the Vienna of *fin de siècle,* which was rich on traditions and ideas but in dire need of sensations and breaks in the boredom of life. Weininger's death was regarded as a consequent conclusion of the argument in his book and its view of human life. This was strengthened the next year 1904, when his unpublished writings, essays, and aphorisms were published as *On Last Things* (*Über die letzten Dinge*).

In his books, Weininger laments modern times as losing the values and grandeur of former eras. This aristocratic view opposes materialism, Marxism, and technology. The individual genius finds no place in this society, and it harbors neither great artists nor great philosophers. Its craving for greatness does not and cannot be satisfied.

The superficial content of *Sex and Character* is repulsive to modern readers; it seems to advocate both anti-Semitic and anti-feministic views, and it has been widely blamed for the occurrence of these trends in the times to come. The main theme of the book centers around the contrast between men and women. This biologically motivated difference is elevated to a general principle for classifying individuals. Everyone is taken to fall on a scale between these extremes. Only the women inclined towards masculinity strive for emancipation. They are thus naturally lesbian and as such should be tolerated but not given any social influence.

The difference between "Man" and "Woman" is thus elevated to principal classification of psychological types. The womanly woman is assumed to lack capability for logical reasoning and initiative. She is only driven by her role as a sexual being, whose only redeeming feature is that she may act as an inspiration and object for the psychological development of the man. Separately, she cannot form judgement and thus concepts of truth and moral carry no meaning for her. Her roles are those of mother or prostitute.

In addition to sexual theories, Weininger analyses the Jewishness of individuals. Also the concept of "Jew" is elevated to a universal psychological term transgressing any borders of race of religion. The Jew is regarded as highly feminine, thus lacking any sense of right or wrong beyond his urge to procreate. Jewishness is a disposition for conformity and mediocrity. Weininger was both Jewish and homosexual, thus presumably a highly feminine type. In this light, his suicide may appear consequent.

In order to understand the impact Weininger's ideas had on Wittgenstein, we must go beyond the superficial presentation above. To this end we must take a more penetrating view of the method and purpose of Weininger's approach.

Weininger was a product of *fin-de-siècle* Vienna. His use of words like feminism and Jewishness cannot be judged from the associations offered by the conflict-filled and violent history of the twentieth century. He picked up the terms from contemporary discourse, and he was addressing contemporary readers. Thus his writings appear to be based on a series of clichés, which, however, were familiar to his audience. According to Wittgenstein, only a young man can be that prejudiced. From the public debate, Weininger inherited a wish to overcome borders and approach the most serious issues in a witty and ironic manner. The master and forerunner here was Karl Kraus and the periodical *die Fackel.*

In order to try to surmise the aims and methods of Weininger, we have to consider his own declaration of the purpose of the book *Sex and Character.* Its subtitle *Eine prinzipielle Untersuchung* is not easily translated into English. A "principal investigation" is certainly wrong. To me the formulation "An investigation into principles" seems to be more adequate. This conforms with the statement by Weininger himself.

> This investigation is not a specialized one but one on principles. It does not despise experimental work, even though the means afforded by laboratories seem to be limited if compared with the accomplishments of analysis by self-observation.

Thus even if the work seems to derive from the progress in psychiatry then so topical, this relation to empirical trends is not the theme of the book. In its stead Weininger presents a tool which he regards as rational and universal for the analysis of the place of the individual mind in the world. This is then applied to the recent scientific findings with disastrous consequences for the author. He lacked insight that his material might be both oversimplified and temporary.

In order to present a method, Weininger introduced a set of ideal concepts taken to have an almost mystical significance.

> Such types not only can be constructed, but must be constructed. As in art so in science, the real purpose is to reach the type, the Platonic Idea.

His prime difference is the Male–Female dichotomy, which is never perfectly materialized in any individual. But in order to judge this, the ideals have to be introduced. They present the *essences* which stratify reality into rational categories. These ideals are not to be found in reality, but they have to be constructed in order to satisfy the heuristic need to pinpoint the differences and similarities between individuals.

In order to illustrate his meaning, Weininger picks out terms from the current discussion and analyzes individuals according to his own terminology of types. Here he purports to use ordinary language only, but his discourse is still rather idiosyncratic. I have already discussed his use of the male–female duality. Another similar conceptual pair is masochism–sadism. Beyond the ordinary meaning conveyed by these words, Weininger identifies a masochist with a person who grasps only totalities and sees the details in an analysis of parts only. The sadist, on the other hand, collects experience of real parts only and sees the totality as symbolizing the collection. The masochist can only relate to himself and is thus not able to relate to woman in contrast to the sadist. But even he is likely to be disappointed in his approach to woman. He will either fall to the level of the woman or regret his adoration of her. A similar pair is "seeker—priest". Here, the seeker strives to define himself whereas the priest has a fixed ego from the start and is thus closed to learning.

Weininger accepts that the human being is an animal and has developed as such. Thus we cannot escape the "animal logic" of our ancestry. This makes it natural for him to adopt a symbolism in terms of animal prototypes. Thus the dog signifies criminality. This is a type lacking a feeling of moral direction, and strives to satisfy only his own cravings. He is the opposite of Nietzsche's *"Übermensch"*. The horse, on the other hand, signifies madness. These terms can also be used to classify genius; the genius is either the reverse of criminality or of madness.

Even if inherited from the animal within us, we may choose the terms in which our person judges his thoughts and actions. Morality cannot be imposed from the outside, each one has to work it out for himself. No absolute Kantian imperative can be found. In this manner we can understand the deep relations Weininger sees between accepting logical statements and ethical issues. They cannot be derived from anything else, because then they would not be fundamental choices. To exist one has to choose a pattern of living. Only within such a framework can moral attitudes be constituted.

Weininger connects the possibility of an ethical attitude to the unidirectionality of time; we have to make our choices for the future; our past actions are beyond amendments. Weininger states:

> The unidirectionality of time is identical with the fact that in its essence a human being is a willing being....
>
> The riddle of time is identical with the riddle of life (though not with the riddle of the world)....
>
> The problem of the unidirectionality of time is the question of the goal of life.

The position of the individual is that he is put into the midst of life and has to find his role in accepting this.

For man to succeed in finding the goal of life, he will have to say "yes" to the world as he finds it; this is an enormous "yes".

This seems to point to an early emergence of the existentialist approach to living in reality.

Influence on Wittgenstein

As we noted, Wittgenstein regarded Weininger as making an "enormous mistake", which could be corrected by negating the whole presentation. So what is there in the approach which is so totally wrong? It cannot be any individual statement, even if Wittgenstein certainly disagreed with many of them. It must be the full premises of the work, its very soul.

Weininger presented his work as an investigation into principles. The illuminations he chose from the contemporary debate were not essential to the main message of the book. So if we are to find what Wittgenstein regarded as false, we have to look at the total approach.

Weininger strived to introduce ideal categories against which to judge the imperfect features of reality. These are to be taken as essences of e.g. "maleness" and "womanhood". This does not necessitate the creation of a formal technical language:

The most extraordinary wisdom is concealed in common speech, a wisdom which reveals itself to a few ardent explorers, but which is usually overlooked by the stupid professional philologists.

Here Wittgenstein would undoubtedly have agreed; in his later writings he constantly fought against the "bewitchment of our intelligence by language". But he was no friend of "essences" and contrasted the impression of an object to itself:

We find ourselves dogmatically conferring on the object properties which only the prototype possesses.

Wittgenstein develops his discourse in a way that differs from Weininger's method. The linguistic tools are only to be used as a means of comparison, never to offer a full picture of the intended object. They are limited to things "which can be said". The formal logical issues have no consequences for how we live our lives. There is knowledge which cannot be transmitted by the formal linguistic means. Examples are how we play games, perform art, or decide about morals. Thus we have to investigate how language is used in its multifarious aspects. These observations can be utilized by the individual to define his own identity. Logical systems can be of no help. There can be no "theory of philosophy" because it would have to be based on something else. The fact that we may state the dualism "A" and "not-A" carries no weight in existential issues.

In modern philosophy, truth for the individual cannot be derived from anything. The ego cannot be captured by rational analysis. Its constituents are multifarious and consist of many different aspects depending on the circumstances. The Cartesian skepticism is not going far enough; the "I think" has to assume an "I" as the subject. But this subject cannot be part of our experience, because it is having the experience. The solipsistic view that the world is my experience does not conflict with the possibility to describe the world realistically. The ego uses a language to describe its experience, and the existence of a language assumes the existence of something. Analogously the world of another may be equally realistic, but if the two worlds are described in the same way we have achieved a scientific consensus.

We are now in the position to try to elucidate the relationship between Weininger and Wittgenstein. First they agree on many things. Both regard ordinary language as the tool to structure reality. Many of the issues and formulations of the later Wittgenstein carry a decidedly Weiningerian stamp.

The difference is, just as we might expect, in the totality of the approach. Weininger introduced ideal essences which allowed him to order the phenomena according to a scale based on similarities and differences. Wittgenstein denies the existence of ideal terms, he wants to catch the linguistic usage by investigating the actual use of sayings as applied to actual situations. Then he finds that meanings cannot be ordered in one-dimensional hierarchies but compose a multitude of intentional actions living their life according to the web of language usage in the appropriate societies. No *a priori* essences can be posited to aid the classification. The structure of reality transcends our mental and linguistic tools.

Thus there is a basic rift between the whole approach of Weininger's work and the views of Wittgenstein. Even if both are motivated by the same public debate and define the same personal problems, no elucidation by Weininger can change the basic method of analysis. Even so, Wittgenstein was attracted by the moral sincerity of Weininger; he had himself exactly the same moral approach to philosophizing. There is no essential difference between the issue of one's sins and acceptance of logic. Both are founded in individual choices based on personal experience. However, the methods the two choose to collect the necessary experience differ profoundly. The difference becomes only more manifest because the same problems are discussed.

We thus see that in spite of their inheritance from the cultural atmosphere of *fin de-siècle* Vienna, Weininger and Wittgenstein reacted completely differently to the challenges offered by the contemporary debate. The former tried to identify from the outside ideal concepts which could be utilized, so to speak to gauge the position of each phenomenon on a well-defined essential scale. Wittgenstein rejects this approach; he works from the inside of human experience and tries to see the actual pragmatic use underlying our communication with the environment including other individuals. Thus he can consider the whole book by Weininger as one single enormous mistake. In denying the whole exposition, he may take exactly the opposite approach, and thus find that this denial makes for a truth which can be seen to follow from Weininger's work. Applications to special cases and different phenomena need not change this; only the total pattern is important. And this can be treated as a *single statement* which can be the object of negation. This way even an "enormous mistake" can give rise to great truth.

As Niels Bohr expresses himself: "A deep truth is one the opposite of which is a deep truth too."

5.2.3. Role of Boltzmann

Ludwig Bolzmann

Boltzmann (1844–1906) is one of the genuine pioneers of modern science. He contributed in many ways to the emergence of modern physics at the turn of the century. His main achievement, however, was in the foundations of statistical mechanics. He was one of the first researchers to understand the profound significance of Maxwell's introduction of probabilistic concepts in gas dynamics. For this he had to accept the reality and significance of the concept of an atom (and consequently a molecule). In spite of the opposition, partly spearheaded by Mach, he pursued his research which forged a solid foundation for thermal equilibrium and the approach to it. The Boltzmann transport equation is a lasting edifice to his memory. His written papers were meticulously produced but unfortunately far from easily accessible to his contemporaries. Consequently the, writings by Gibbs, which are in many ways more formal, became known sooner to the research community.

Boltzmann had a thorough education in philosophy even if he did not set out to emphasize this in his research. However, the intellectual atmosphere in Austria compelled him to look into the conceptual foundations of his work. This led to his appointment to the chair in natural philosophy when Mach retired. In this capacity, he wrote a series of popular articles where he lays out his views [14]. Here all quotations of Boltzmann are from this collection.

As a physicist, Boltzmann was an ontological realist; he considered physical objects as real. The theories proposed to describe them he regarded as free creations of the human mind. They are validated by their success only; the right theories survive, the useless ones disappear. Here Boltzmann was strongly influenced by the Darwinian model of biological evolution, and he applied it to mental phenomena too. Consequently his views were closely similar to those of the contemporary *pragmatic* movement. However, he does not seem to have been aware of this.

Boltzmann's life ended sadly in suicide in 1906. This was presumably not caused by the opposition to his physics, there he was confident enough, but rather by a general depression possibly precipitated by his failing eyesight, which threatened to blind him in the not too distant future.

Here our interest in Boltzmann derives from his being a forerunner of Wittgenstein. He is listed as one of the three physicists influencing Wittgenstein who, it is known, had intended to study with Boltzmann—a plan that was thwarted by the suicide.

It is thus certain that Wittgenstein had read Boltzmann's essays. However, it is surprising how close to Wittgenstein's ideas one comes in reading Boltzmann. Both the demarcation of meaningful propositions from metaphysics as well as the interdict against meaningless statements are found in Boltzmann's works.

Surprisingly enough, also the roots of scientific positivism emerge from the texts. It seems unlikely that Bohr knew the writings of Boltzmann, but their general tenets are clearly discernible in Bohr's philosophy. Such ideas thus seemed to linger in the scientific community.

Empiricism

Boltzmann as a scientist obviously had to take a realistic view of matter. In particular, his use of the concept of "atom" was granted evident existence: "It would, of course, be absurd to prove or disprove the objective existence of matter."

He realized that knowledge can only build on a shared platform of agreed concepts (cf. Wittgenstein's *On Certainty*).

> The wealth of experience as well as words and ideas used for denoting it, must be presumed as known if we are to understand each other at all, and that we cannot define everything.
>
> The ideas of the thinker himself cannot be left out of the world picture if the world picture is not to vanish completely.

Bohr would have stood behind this, if he had known Boltzmann's writings, which is unlikely. Theories have to be built consistently and according to the progress of science:

> Progress in thinking must be sought by eliminating such forms of inference and concepts which do not inform but mislead and even entangle us in contradictions if we apply modes of thought transferred to cases where they do not fit. Actions that are followed by things we desire and ideas that are correct and economical as well, we denote as correct.

Those theories which work are selected by a Darwinian process (cf. Popper's *falsification*):

> Of opinions sufficiently often confirmed by experience, we say that they are certain and that what they express belongs to our knowledge.
>
> What leads to correct deeds is true.
>
> What then will be the position of the so-called laws of thought? In the light of Darwin's theory, they will be nothing but inherited habits of thought.

Here we are very close to a *pragmatism* of James's brand. As pointed out above, it seems, however, unlikely that Bolzmann knew about James.

Human mind

The insight that all scientific activity plays on a mental arena will require some elaboration.

It is a peculiar drive of the human spirit to make itself such a picture and to adapt it to the external world. The immediate elaboration and constant perfection of this picture is then the chief task of theory. The more abstract the theoretical investigation, the more powerful it becomes.

Words and therefore concepts we can form as we wish.

The ability to form mental constructs seems to like stretching beyond what is motivated by the actually experienced:

The drive towards philosophizing seems ineradicably innate.

Metaphysics seems to cast an irresistible spell on the human mind.

The eminent importance of the right method explains why men soon started to think not just about things but also about the method of thinking itself; thus arose the so-called theory of knowledge, which in spite of certain tang of old-style metaphysics now discredited, is highly important to science.

The mental processes tend to acquire a life of their own.

It is indeed true that we could have no experience if certain forms of linking perception, that is forms of thought, were not innate to us.

However nothing seems less founded than an inference from this *a priori* to absolute certainty and infallibility.

The simplest things the philosopher finds puzzling. These, however, are nothing but the inappropriate and mistaken mental re-shaping of what is given, which cannot contain any contradictions.

It is indeed so that the forms of thought that have become habitual are not quite adapted to the abstract problems of philosophy.

Niels Bohr always was very careful not to state more about a physical phenomenon than he could express in "habitual forms of thought", i.e. common language:

Since we cannot clearly express such complicated conditions by words, whether written, spoken or thought, we must collocate words to lend the most fitting expression of the given.

When these concepts and pictures are applied to metaphysical questions, we seem to approach what Wittgenstein termed *Unsagbares*.

In this way a deception of the intellect supervenes with philosophical problems. Our subjective feeling, that is fine and lofty which advances and raises our species, objectively these concepts do not exist.

Theories and reality

In order to communicate experience between individuals, Niels Bohr always stressed the need for unambiguous descriptions. Boltzmann writes:

If I am to make myself understood, I must adopt a language in which all exist on the same footing ("objectively"). This adherence to the language of others which is given me in experience (because learnt) I call the objective point of view, in contrast to the subjective one.

Boltzmann, as an empirical physicist, affirms that our scientific progress provides knowledge about reality:

Theory is only a mental picture of phenomena, related to them as sign is to designatum.

Hertz is not looking for a satisfactory mechanical derivation of these fundamental equations. He points out that the correctness of all these equations is not the few experiments from which they are usually derived, but rather their subsequent agreement with almost all known facts. As we have these equations, we better write them down without derivation, compare them with phenomena and regard constant agreement as the best proof that the equations are correct.

Here we encounter the "context of discovery" as contrasted to "context of verification". Boltzmann realizes that even the best theory is underdetermined by the empirical evidence, we cannot prove a theory "right":

The assertion that a given theory is the only correct one can only express the subjective conviction that there could not be another equally simple and fitting image.

Maxwell warned against regarding a particular view of nature the only correct one merely because a series of consequences flowing from it has been confirmed by experience. He gives many examples how a group of phenomena can be explained in two totally different ways, both modes of explanation representing the facts equally well.

The new approach compensates the abandonment of complete congruence with nature by the correspondingly more striking appearance of points of similarity.

Our aim will not be to establish the truth or falsehood of one or the other world picture, but we shall ask whether either is appropriate for this or that purpose while we allow both pictures to continue alongside each other.

The only justification for a world picture is its agreement with the facts.

Limits of representations

As even the best theories are but pictures called up by our minds, we should not be over-confident in their validity; we cannot derive the world from abstract structures:

We must not aspire to derive nature from our concepts, but must adapt the latter to the former. It may happen to the mathematician, that he takes their mutual relations as the really existing and turns away from the real world.

It remains the ultimate enigma that our theories may serve to map reality to the extent it does:

The most ordinary things are to the philosophy the source of insoluble puzzles.

I always had a nightmarish feeling that it was an unresolvable puzzle how I could exist at all, or that the world could and why it should be precisely as it was rather than otherwise.

Wittgenstein shows that propositions which can be proved logically to be true, are tautologies telling nothing about physical reality.

All ideas of general phenomenology show themselves to be pictures of phenomena. No more must mathematicians confuse their formulae with truth, or they will be similarly blinded.

I absorbed Darwin's theory and saw from it that it was really a mistake to ask in such a way that the question can not be answered.

Boltzmann goes on to point out the bounds on human capability and formal systems.

Asking if life has a value has no sense; we can only ask how life is to be promoted. The valuable is simply what promotes life.

Laws of thought overshoot the mark...

likewise if we ask why the world exists at all etc. The simplest preconditions and the laws of thought can at best be described.

He goes on to envisage the proper role and task of philosophy:

I regard the existence of these questions and problems themselves an illusion.

If therefore philosophy were to succeed in creating a system such that in all cases mentioned it stood out clearly when a question is not justified so that the drive to ask it would gradually die away, we should at one stroke have resolved the most obscure riddles and philosophy would become worthy of the name of queen of science.

Philosophy of the future is to formulate the fundamental concepts in such a way that we obtain as precise instructions as possible how to intervene in the world of phenomena.

Boltzmann thus formulates the conclusion written by Wittgenstein in the *Tractatus* (6.53):

The correct method in philosophy would really be the following: to say nothing except what can be said, i.e. propositions about natural science - something that has nothing to do with philosophy - and then whenever someone else wanted to say something metaphysical, to demonstrate to him that he had failed to give a meaning to certain signs in his propositions.

Except for the overly optimistic view of scientific knowledge, this appears to derive directly from Boltzmann's writings.

5.2.4. Interview with von Wright

Georg Henrik von Wright (1916–2003) was one of the few persons who came close to the enigmatic Wittgenstein. They were close friends, and when Wittgenstein retired from his professorship in Cambridge, von Wright was called to be his successor, a position he held from 1948 to 1951, when he returned to Finland.

Evidence of Wittgenstein's appreciation of von Wright was the fact that he elected as the executors of his considerable *Nachlass* the following three: Elizabeth Anscombe, Rhus Rhees, and G. H. von Wright. This material consisted of huge amounts of notes and scattered comments, much of which has been screened and published over the years. Von Wright and his collaborators have played a major role in this process. Added to Wittgenstein's personal relation with von Wright was, no doubt, the fact that he was not pursuing philosophy according to Wittgenstein's method, and thus von Wright was immune to the temptation to try to emulate a method unreachable for ordinary man.

On January 4, 2003, I had the opportunity to see von Wright and ask him some questions about Wittgenstein. Here I reproduce my questions, and the response according to my notes after the meeting. Some overlap with earlier parts of my text is inevitable. However, I let it stand as a testimony by a source close to Wittgenstein.

Question: What had Wittgenstein read, and what was he possibly influenced by?

G. H. v. Wright: He had read parts of James. Much of his psychological writing was response to this. The psychological fragments are late. They represent a novel direction, which had emerged but not reached a definite form. W. thought that something new was emerging. Not yet ready.

W. had read Spengler and Tolstoy, but also Dostoevsky. He knew Nietzsche but had not read him systematically. From an existentialistic point of view, W. has been compared to Heidegger.

W. knew Freud and had done some reading. Freud knew W.'s sister Margarete (Stonborough), who helped him escape from Austria.

W. claimed that Freud reminded him of Chaplin, who did not see the hazards (truth?) but avoided them by accident. Freud's theory was good, but he could not express it in words. According to W., Chaplin too should have refrained from turning to the use of words. The silent movies were the best ones. W. took a negative attitude to the "Dictator".

According to W., Sartre was like a fish on land. He was snapping for air.

v. W. admitted that W. took a critical attitude to most writings he read. Philosophy is unsatisfactory (including his own) and it ought to be possible to eliminate it.

Question: What did W. know about physics, especially modern physics?

G. H. v. Wright: W. had read Hertz, Mach, and Boltzmann. He apparently liked Boltzmann, who thought that Hertz expressed himself badly. W. had said that he did not like electrodynamics. As far as v. W. knew, W. never commented on atomic physics. He appeared not to be aware of the discussion about modern physics, which emerged in the 1920s.

Answering the question, v. W. verified that W. was innocent concerning the discussion about physics. He is thus not affected by the discussion. He also never talked about relativity.

W. never mentioned Heisenberg's uncertainty nor complementarity. Here v. W. declared that Bohr is unclear in his fundamental philosophical attitude.

Question: What did W. think about Gödel, theories of truth, and many-valued logic?

G. H. v. Wright: There exists a "paraconsistent logic" by Newton da Costa, which attempts to include "contradiction" into a logical system. v. W. himself has written about "truth logic".

W. discusses probability in *the Tractatus.* v. W. has written about this. The name of the work was suggested by Moore; this is how the Spinoza connection arose. It is improbable that W. had read Spinoza. W. did not read much. He took a book, browsed through it for things he could assimilate.

This concludes my notes from our discussions. Much of it is known, but I have chosen to include everything I put into my own notes.

The interview confirms certain opinions expressed elsewhere in this work: Wittgenstein knew only mechanics and nothing about modern physics. His attitudes were strongly influenced by Boltzmann. Wittgenstein did not read much even in philosophy.

Georg Henrik von Wright passed away on June 16, 2003, a few months after my interview with him.

5.3. Stages of Wittgenstein's philosophy

5.3.1. The early Wittgenstein

The early thinking of Wittgenstein is collected in his work *Tractatus Logico-Philosophicus*, which he finished as a prisoner of war at Monte Cassino in Italy. He managed to get the manuscript to Gottlob Frege, Bertrand Russell and Paul Engelmann. The first two were the leading representatives of mathematical philosophy at the time. Engelmann, on the other hand, was an Austrian architect, who was well versed with the cultural atmosphere and debate of Vienna. He was to collaborate with Wittgenstein in designing the house for Wittgenstein's sister in Vienna. He has published his correspondence with Wittgenstein, elucidating the ideas and developments behind the philosophy of Wittgenstein.

Frege did not understand the point of the work, and Russell wrote a preface, which according to Wittgenstein misses its point. The work, however, needed to be published. The tortuous road ended with final publication in 1921 by the journal *Annalen der Naturphilosophie.* Wittgenstein was never satisfied with this printing, and he also disliked the fact that Russell's preface constituted the ultimate motivation for its acceptance. A more satisfactory version, with English translation opposite the original text appeared in 1922. At this time, the work got its final Latin title at the suggestion of Moore. The inspiration was, of course, Spinoza's *Tractatus Theologico-Politicus.* The *Tractatus* was to remain Wittgenstein's only philosophical book to be published during his lifetime.

This is not the place for a detailed analysis of the argument in the *Tractatus*; generations of philosophy researchers have devoted themselves to such analysis. However, in order to proceed, I need to summarize the main structure and its common perception. Any summary of the argument is likely to be declared misguided by the experts. This is unavoidable, because they do not provide a consensus view themselves. Especially, a brief review like mine is bound to be misleading. For this I may be forgiven, I am here in the company of my betters.

The work is presented in a numerically structured series of aphorisms, where deeper decimalized sentences are subordinate to the higher ones. The single digit sentences may thus be regarded as a summary of the structure. The logical argument is, however, interwoven all through the text, and the themes extend through the parts, combining to forward the structural pattern.

1. The world is all that is the case.
2. What is the case—a fact—is the existence of states of affairs.
3. A logical picture of a fact is a thought.
4. A thought is a proposition with a sense.
5. A proposition is a truth function of elementary propositions (An elementary proposition is a truth function of itself.)
6. The general form of a truth function is $[\bar{p},\bar{\xi},N(\bar{\xi})]$.
   This is the general form of a proposition.
7. What we cannot speak about we must pass over in silence.

An attempt to interpret the flow of thinking indicated would be:

#1 just defines what we may consider, the ubiquitous ambience where we live and act.
#2 introduces those elements of the world description which we can use to obtain a consistent map of experienced reality
#3 uses the description developed to connect the formal logical framework to the mental ideas experienced as thought.
#4 introduces the meaning of the logical framework by asking for its sense.
#5 analyzes the formal structure as to its logical relation to truth.
#6 selects what can be logically ascertained and what we must exclude from our analysis because it is ineffable and only shows itself.
#7 gives the main conclusion and the edict not to formulate questions lacking meaning.

Superficially this seems to be a treatise on logic. The first point defines "the world". In the following I will use the term "world" with the current philosophical connotations; no ambiguity is assumed to adhere to it. Wittgenstein makes its description consist of a series of independent binary propositions. It seems clear that he also regards the fact that some case does not hold as information in need of being stated explicitly. To determine the world, all relevant propositions have to be listed. The facts of Wittgenstein have to be independent and form a complete description. Much discussion has been concerned with the Wittgensteinian concept of a fact. It seems reasonable to assume that he considers relations things that need to be affirmed or denied. The approach can be taken to formalize the Leibnitz concept of a "possible world", which has played a central role in modern semantic discussions. It has also been pointed out that this "logical space" of Wittgenstein reminds us of Boltzmann's state space of the configuration of a physical system.

The second point stresses that the configuration is the essential element of the world. As he aims at mirroring its logical structure, he presents "states of affairs" and the objects, things carrying these interconnections, may not need to be fixed in advance. The symbolic representations are only models, "pictures" of the states of affairs. It represents them and this gives the picture its sense. The ideas may originate in the logical atomistic position that the world has to be constructed from "sense data", which played a central role in the later developments. However, Wittgenstein never refers to the sense data as the elementary propositions from which he constructs the complex propositions utilized in his world making.

Point 3 presents a central idea of the work. The vision of the world conveyed by our making decriptions like a "picture". It relates to our thoughts in such a way that we can envisage a state-of-affairs only in the form of a picture. This, of course, does not comprise only visual images but all logically conceivable structures. A picture can only be constituted as a description of the world if its form is such that it obeys the correct logical syntax. Only such states-of-affairs can be the objects of our thinking. An analogy with a map is, perhaps, more relevant, as the contact with reality is merely a relational similarity between the representation and the experience. Erik Stenius especially has discussed the possibility of a world description based entirely on binary statements acquiring only the values "true" or "false" [93]. The irony is that the present trend in science to define all knowledge as information also rests on the possibility of binary descriptions.

In point 4, Wittgenstein introduces the limits to thinking which the picture theory presents. We can comprehend only a complex proposition whose truth value derives from the truths of its constituent elementary propositions. All situations with sense can be described in human language. A proposition which is always true, a tautology, cannot describe the world, it puts no conditions on its possible being. Likewise a contradiction describes no world, because it will never prevail. The complexity of human life, however, makes it impossible to analyze every expression as to its logical structure. From this Wittgenstein concludes that thoughts may describe sense in reality without the logical language being able to convey this; it may show itself but be beyond expression in the world description.

Point 5 states that logical propositions derive truth value only from the truths of its various elementary propositions. It seems that Wittgenstein assumes the truth values of these as unproblematic. Our everyday discourse, and also scientific descriptions, are fully correct as they stand. Maybe their truth value shows itself; this may be the ultimate criterion for the selection of primary propositions. If a statement cannot be dissected into such elements, it may still show something, but its truth cannot be stated. Wittgenstein concludes that the limit of one's language is the limit of one's world. To make a statement beyond this limit is not warranted. His aim seems to be to eliminate statements of the metaphysical kind. There can be no necessarily true propositions except tautologies, and these do not pertain to the world. A metaphysical statement may be formally correctly formed, but after analysis we may find that some of its constituents are not sensible elementary propositions. Hence the whole statement lacks sense.

Point 6 appears to be fully logical only, but it leads to the final conclusions of the work. It is a formal expression, the meaning of which has been explained by Russell, in the preface of the *Tractatus,* and by Mounce [94]. It appears to state that given any set of elementary propositions, true by fiat, and any set of propositions obtained by allowed operations, we can obtain a new proposition by further applications of the formal operations. The method is well known in mathematics as recursive definitions, and its use in logical proof theory is obvious.

Wittgenstein concludes that all logically true sets of propositions can only be tautologies. They derive from a set of manipulations, and they can carry no significance for reality. All proofs are only expedients to bring out into the open what is inherent at the starting point. Thus if logic necessity is applied to reality, it has to hold in all possible worlds and gives no information about the one we live in. At the time Wittgenstein believed that there could exist a formal procedure to write down all logically obtainable propositions; a computer should be able to do it. Thus it was a major shock to the following generation of logicians, when Kurt Gödel in 1930 presented his incompleteness theorem: In every formally defined logical system, statements can be formed whose truth value cannot be derived within the system. Furthermore, he demonstrated that some such statements can be seen to hold. The proof rested on the consistency of arithmetic, which has not been proved.

The logic is not absolutely prescribed but determines the structure of possible pictures of the world. Mathematics is not in itself part of this picture language. If the matter-of-facts concerning the world is all we can state in our language, it follows that no "higher values" can be part of our world. This is brought out in point 6. The sense of the world cannot be found in it. No values can be based on our world. Ethics and esthetics must be beyond our language. Religious propositions are not a feature of the world; God does not reveal himself in the world.

It is not *how* things are in the world that is mystical but *that* it exists (*Tractatus* 6.44). Feeling the world as a limited whole - it is this that is mystical (*Tractatus* 6.45). When the answer cannot be put into words, neither can the question be put into words (*Tractatus* 6.5). We feel that when *all possible* scientific questions have been answered, the problems of life remain completely untouched. Of course there are then no questions left, and this itself is the answer (*Tractatus* 6.52) There are, indeed, things that cannot be put into words. They *make themselves manifest Tractatus.*(6.522)

From this Wittgenstein derives conclusion 7 of the work: "What we cannot speak about we must pass over in silence."

There are many books published on the argument and conclusion of the *Tractatus*. I do not claim to know many of them, and any expert will be able to find the shortcomings of my résumé. It may still serve as a reference for the use I will make of the text later.

The work was not understood by many, and its aim and method has become clearer only over the years. One reason is that the text actually consists of two intermingled discourses. In one of his first attempts to get his manuscript published, Wittgenstein wrote to the prospective publisher Ludwig von Fricker:

> I wanted to write that my work consists of two parts: the one which is here, and of everything I have *not* written. And precisely this second part is the important one. For the Ethical is delimited from within, as it were, by my book; and I'm convinced that, *strictly* speaking, it can ONLY be delimited in this way.

By making a formal scaffolding for all that can be stated and founding the boundary of this, Wittgenstein proclaimed that all "higher values" cannot be subjected to logical discourse. Allan Janik and Stephen Toulmin [21] have convincingly argued that here we encounter a contribution to the discussions in the cafés of Vienna as described above. Moral considerations and beauty versus utility had been at the center of this discussion. In the satirical writings by Kraus, the language had been shown to say what could not be said.

The possibilities and limitations of language had been at the center of the discourse. Already in 1901, the journalist-turned-philosopher Fritz Mautner had published his extensive work *Contributions to a Critique of Language.* Inspired by Schopenhauer, he investigated the use of language, and discussed both its role as a pictorial representation and its role as a tool of acting and living humans. Wittgenstein acknowledges the inspiration from Mautner in his typically aloof manner (*Tractatus* 4.0031): "All philosophy is a 'critique of language' (though not in Mauthner's sense)."

Combined with the earlier presented discussion of the role of science, the analysis of language in Vienna was prepared for Wittgenstein. In the first decades, when he wrote his treatise, he took the logical tools and formulations from Frege and Russell, but the problems he set out to solve were formulated on the home range. His logical apparatus was based on them but also on a penetrating criticism of their approach. The things he did not write about, he took to be so essential to the human situation, that what he had proved implies that they could not be discussed in language. This was taken to be a positive statement of their character. Every reader was supposed to realize this transcendent truth. Only art can, in its most engaged form, give glimpses of the true human predicament.

Thus Wittgenstein regarded his work as finished. He had shown that the world consists of what one can say, that only tautological expressions can be necessary, and that the many other issues showing themselves in the human situation, cannot be discussed in language. "Darüber muss man zweigen." There was to be no further need for philosophy, so Wittgenstein withdrew from the field as we have seen in his biography above.

But the *Tractatus* consists mainly of a series of purely logical arguments. In order to state his world view Wittgenstein utilizes all the formal resources he found in Frege and Russell. He is, maybe, not so much a pupil as an independent laborer borrowing his tools from the masters of logic. Thus the work was seen as a contribution to formal logistics; its aim was to find the perfect language of philosophical discourse. Wittgenstein had no such aim in mind, but Russell seems to have got this impression when writing his introduction. But Russell, presumably, did not know the discussions in the cafés of Vienna.

In order to formulate the linguistic possibilities to describe the world, Wittgenstein utilized the formal methods of logic. He was thus, in a way, providing the framework for the models Hertz had envisaged as the aim of physics too. All terms have to be used in their proper way, the propositions have to be joined in meaningful ways, and only complete statements can be set up against reality and found to be valid or not, true or false. Only those combinations which are relevant for reality can be seen as having sense. We cannot test the validity of incorrectly constructed pictures, and our language, if correctly used, cannot describe an impossible world.

In the course of this activity, Wittgenstein introduced a terse language and formal constructions to test if statements make sense (truth tables). The text was formally compact and his statements were aphoristic instead of dialectic. The work thus offered a challenge to the logistic readership, which was eagerly taken up by the formal investigations into semantic and epistemological discussions of the concept of truth. Wittgenstein's conclusion was that in a language we cannot tell what cannot be said in that language. For a pure rationalist, this seems to be something of a triviality. We have, however, seen that Wittgenstein was claiming more: The language can be used to show when we approach this border; the limitations of language exclude all influences and motivations steering our lives. As we exist and act in the world, the conditions of existence show themselves but it is beyond human capacity to formulate them. These subtleties were lost on the pure logicians but were in concord with the later developments of Continental philosophy.

Finishing this work, Wittgenstein believed he had disposed of the need for philosophizing. All possible questions were either scientific or empty; the former can but be decided by science, the latter can be disposed of as meaningless (*Tractatus* 6.53).

5.3.2. Wittgenstein's influence

The ground was fertile for the purely logistic aspects of philosophy, and Wittgenstein became a model for these developments at first. Frege and Russell had kindled the interest in the foundations of mathematics, and together with Whitehead, Russell was claiming that logic proved the firm foundation.

A group of rationally oriented scientists used to meet at the cafés in Vienna to discuss the ideas inherited from Mach and Hertz. Among others these consisted of the mathematician Hans Hahn (1880–1934), physicist Philipp Frank (1884–1966), and economist Otto Neurath (1882–1945). These formed the precursor to the Vienna Circle still to be born. They were eventually to exert their influence over the German-speaking realm, Frank in Prague, Neurath in Munich, Hahn in Vienna, where Neurath returned after a brief political adventure in Bavaria. When the chair of natural philosophy in Vienna was filled in 1922, the newcomer was Moritz Schlick (1882–1936), who was a physicist turned philosopher. He became the founder of the Vienna Circle starting in 1924, and he remained very influential up to the time when he was murdered by a mentally disturbed student. Among other early participants in the discussions, we may mention Viktor Kraft (1880–1975) and Friedrich Waismann (1896–1959). They were joined by the mathematician Kurt Gödel (1906–78) and Rudolf Carnap (1891–1970). These formed the center for the group referring to itself as the Vienna Circle in a pamphlet from 1929.

The aim of the activity was to unify all branches of human knowledge, and derive all possible factual relations from unified and simple "protocol statements", the elementary propositions of Wittgenstein or the atomic facts of Russell. The spirit is captured in Carnap's *Logische Aufbau der Welt* from 1928. Following the British empiricist, the original intention was to build the theory upon "sense impressions". The endeavor did not lead to final satisfaction, partly because the elementary statements could not be agreed upon. The whole concept of meaning and nonsense was questioned by Carnap in *Testability and Meaning* from 1937–8. Similar conclusions are presented by Quine in *Two Dogmas of Empiricism* from 1953.

The activity in Vienna became widely known. Foreign participants visited the group, learned their approach to philosophy, and brought home their ideas. Thus they were visited by Americans Ernest Nagel (1901–85) and W. v. O. Quine (1908–2000), the Brit A. J. Ayer (1910–89), and the Nordic researchers A. Næss (1912–2009) and E. Kaila (1890–1958). When Hitler took over power in Germany in 1933, all scientists of Jewish origin got into difficulties. The discussion circle in Vienna was closed down, and many of its representatives were dispersed. Neurath was visiting in Moscow and chose not to return but settled in the Hague. When the Germans conquered Holland in 1940, Neurath saved himself by fleeing to England. Carnap moved from Prague to the US in 1935. Popper did not belong to the Circle, but carried his brand of philosophy first to New Zealand and then to London. Waismann was received in Oxford. Gödel left for the US as late as in 1940. Thus the political upheaval in Europe dispersed the Viennese tradition in philosophy around the world, and the novel logical approach to conceptual investigations became known as a purely Anglo-Saxon movement in spite of its many Austrian roots. Thus the cultural background of Wittgenstein's endeavor was misunderstood and forgotten.

5.3.3. The later Wittgenstein

Investigating language

In addition to the *Tractatus*, the only work Wittgenstein prepared for publication is the *Philosophical Investigations*, which was published posthumously in 1968. This work, however, is based on an extensive body of work in progress. Much of this has later been published. To get a general view of "the late Wittgenstein" we may take his final approach to philosophy to be represented by the *Philosophical Investigations*. As his views were constantly changing and developing, this is far from the full truth, but it may remain a starting point for the discussion of his later attitudes. In the following, the abbreviation *PI* refers to this publication [77].

The aim of Wittgenstein's *Tractatus* is to construct a solid theoretical structure for language, including a demarcation of its limits. *Philosophical Investigations* denies this possibility and is hence often regarded as radically different from the earlier thoughts. This is an oversimplification, because there is a clear line of development in his documented works. But the differences emerging form a significant part of the Wittgensteinian approach to philosophy: It does not aim at final systems or established results, but presents methods and offers doubts related to the use and misuse of language in different situations. This lack of formalized method makes *Philosophical Investigations* look different from customary philosophical writings and causes difficulties in extracting the insights presented. It is easy to misunderstand Wittgenstein's aims; indeed, one is bound to miss many of the subtle points. Thus in trying to summarize the contents of *Philosophical Investigations*, I am aware of the shortcomings of my text. These derive partly from the convolved presentation by Wittgenstein and partly from my limited understanding thereof.

Wittgenstein chooses St Augustine as a representative of the idea that language is a mirror of the speaker's inner processes. Wittgenstein strongly disapproves of the posited existence of the inner person who thinks, plans, and speaks; this invites metaphysical speculations about *a priori* relations between words and external objects. As a model of language acquisition in a child, Augustine's description is clearly both dated and inadequate. It has been suggested that Wittgenstein chose this historical sample in order to avoid pointing out any contemporary philosophic movement. Be that as it may, his criticism could also have been carried by Plato. In the work *Cratylus* [95] there is a discussion of the role of language, which is exceptional in Plato's writings:

The name is an instrument of teaching and distinguishing natures. . . .

The thing may be named, and described, as long as the general character of the thing you are describing, is retained. . . .

But we may admit so much that the knowledge of things is not to be derived from names. . . .

What other ways can there be of knowing them [the names] except the true and natural way, through their affinities, when they are akin to each other, and through themselves?

Socrates' adversaries in this dialogue represent various metaphysical views of language. Cratylus believs that names are natural and derive from the inner truth of the named. His opponent, Hermogenes, claims that names are totally conventional, and any name chosen is as correct as any other one. Through his dialectical analysis, Socrates demonstrates the shortcomings of both views and presents his own synthesis of the argument about the art of naming.

Foreshadowing Wittgenstein, Socrates argues that language is arbitrary but acquires meaning only by use and communication within a society of speakers with a common language.

To repeat—naming is something like attaching a label to a thing. One can say that this is preparatory to the use of a word. But what is it preparation for?(PI I-26)

Language is, however, more complex than this description suggests. Questions like "What is time?" and "What is meaning?" appear to be clear when no answer is solicited, but baffle us when we try to formulate an answer. They are not scientific queries allowing a final elucidation of some definite answers. This violates our inherited thinking about the role of language.

Philosophy is a battle against the bewitchment of our intelligence by means of language. (PI I-109)

The dissection of language Wittgenstein calls a grammatical investigation. For him this is an exploration of words as they are actually used. No explanation is possible or needed but only the description of practice. No stipulation of meaning is imposed on the observations.

Philosophy leaves everything as it is.(PI II-124)

Our intuitive feeling that language must be based on some underlying theoretical structure is without ground. A word is only what it is, and it carries with it intimately its own use.

Since everything lies open to view there is nothing to explain.(PI I-126)

Wittgenstein's concept of grammar thus covers both syntactical and semantic aspects. Wittgenstein himself, however, would presumably deny the existence of these theoretical categories.

A clear analysis leads to perspicuous representations.(PI I-122)

When this is achieved, no philosophical questions remain; philosophy has come to an end.

Once you know what the word stands for,you understand it, you know its whole use. (PI I-264)

By applying the dialectical method of Wittgenstein to one conceptual situation after another, we do not learn a method but adjust our thinking from regarding the customary as obvious. This is like therapy, it turns our thinking.

Language-in-use can only function within a society sharing the same pattern of acting on verbal cues. There must be a communication nexus which allows sensible reactions to verbal utterances. This cannot be global, comprising everything experienced and possible to be communicated. The activity is confined to the nexus and has its life there. On being addressed, what you conclude is not the issue but only what you should do. The same individual may well be part of different nexuses, playing the violin and designing an engine do not necessarily reside in the same practices. Such reality-adapted communication networks Wittgenstein calls "forms of life".

Wittgenstein considers the user's language mastery as a toolbox at his disposal. There are many kinds of tools adapted to many kinds of tasks. They aim at various different goals and attain them more or less adequately. There can be no universal toolbox, and hence the question of their completeness does not arise. Tools may be very differently adapted to different use. They are, however, related by a "family resemblance"; there is a possibility to combine them as tools not by a common feature of all but tracing through the box from one tool to a next one.

A toolbox adapted to its communication needs Wittgenstein terms a "language game". Their number and extension are not given. They rise and wane with the communication needs of the society that nurtures them.

There are countless kinds [of propositions]: countless different kinds of use of what we call "symbols", "words", "sentences". And the multiplicity is not something fixed, given once and for all; but new kinds of languages, language games as we may say, come into existence, and others become obsolete and get forgotten.(PI 23)

But even language games have to be learned. Wittgenstein draws attention to the difficulty of how to get started. Individual words can be pointed out, ostensive definitions, but this works only against a background fabric of practice. Sentences like: "This is red", "This is number 5", and "This is the king" work only if the concepts of color, number, or rules of chess are presupposed. How did we ever get started?

The private and the environment

What kind of ontology does Wittgenstein support? He dissects talking but what can we be talking about? He does seem to question that what is said is motivated by our experience, by what we "see".

"What I really see must surely be what is produced in me by the influence of the object."—Then what is produced in me is a sort of copy, something that in its turn can be looked at, can be before one, almost something like a materialization.(PI II-199)

This posits a mental image, the existence of which Wittgenstein cannot approve. However:

When I imagine something, or even actually see objects, I have got something which my neighbor has not.(PI I-398)

Wittgenstein denies the sense of this claim to internal knowledge.

Might I not ask: In what sense have you got what you are talking about and saying that only you have got it? Do you possess it? You do not even see it. Must you not really say that no one has got it? And this too is clear: if as a matter of logic you exclude other people's having something, it loses its sense to say that you have it. (PI I-398)

In connection with the term "pain" Wittgentein elucidates his view even more explicitly:

Imagine a person whose memory could not retain what the word "pain" meant—but nevertheless used the word in a way fitting in with the usual symptoms and presuppositions of "pain". Here I would like to say: a wheel that can be turned though nothing else moves with it, is not part of the mechanism.(PI I-271)

Mental constructs that are not (logically) necessary for the observed phenomenon are not to be part of the philosophical understanding.

Wittgenstein then goes on to discuss the drawing of a duck-rabbit. This is a simple example of the ambiguous pictures that psychologists have developed: you see this and suddenly you see that. Such an instantaneous gestalt switch is not confined to visual images: you recognize a bit of text as coming from a well-known play, you see the proof of a mathematical theorem, or you recognize a sequence of notes as deriving from an opera. In a moment your experience of a stimulus changes; what happened in the mental state if such an entity makes any sense?

Someone utters the word "rabbit". What can we conclude?

If you are looking at an object, you need not think of it: but if you are having the visual experience expressed by the exclamation, you are also thinking of what you see.(PI II-197)

Thus observing a vocal uttering, there is no unique conclusion about the accompanying mental event.

The concept of a representation of what is seen, like that of a copy, is very elastic, and so together with it the concept of what is seen. The two are intimately connected. (Which is not to say that they are alike.) (PI II-198)

Wittgenstein thus entertains a dual attitude to mental constructs: They cannot be meaningfully denied but we have no clue how to derive their functioning from observed (verbal) behavior.

Role of mind

In spite of all remaining questions, languages exist and they are used. Thus we have a phenomenon well established and open to observation and description in its entire multifarious splendor. No hidden theoretical background is needed; it is all openly displayed for our scrutiny. Wittgenstein's grammatical analysis always aims at revealing the simple nature of a phenomenon.

Wittgenstein devotes consequently a great effort to unravel our response to words. How do we know how to act on a proposal and how do we know how to continue a sequence (i.e. follow a rule)? What private language do we consult to respond properly to linguistic clues?

What kind of mental state do we enter when we grasp the meaning of a rule? According to Wittgenstein, this question violates the grammar of the words used. The understanding is simply a term capturing the ability to act appropriately in the light of previous experience within the appropriate life form. The very meaning of a rule is a social concept defined only in the context of a life form.

And also "obeying a rule" is a practice. And to think one is obeying a rule is not to obey a rule. Hence it is not possible to obey a rule "privately": otherwise thinking one was obeying a rule would be the same thing as obeying it. (PI I-202)

Philosophy of psychology

From the concept of following a private rule, Wittgenstein proceeds to what is known as his private language argument. This discussion forms the mental basis for a philosophy of psychology. Wittgenstein read and was influenced by the work of William James on psychology, and much of his psychological writings constitute a commentary and criticism of the work by James.

James considered consciousness to derive from mental states, which could only be investigated by introspection. This would be formulated in internal language not accessible except by such introspection. James, however, issues a stern warning [55]:

The mental state is aware of itself only from within: it grasps what we call its own content, and nothing more. The psychologist, on the contrary, is aware of it from without, and knows its relations with all sorts of other things. What the thought sees is only its own object; what the psychologist sees is the thought's object, plus the thought itself, plus possibly the rest of the world. We must be very careful therefore, in discussing a state of mind from the psychologist's point of view, to avoid foisting into its own ken matters that are only there for ours.

Wittgenstein takes up this challenge: He does not explicitly deny the existence of introspection, but he cannot accept this as a tool to gain knowledge of any possible mental state. According to his philosophical inclination, no mental entities can be posited on the basis of introspection. Thus only behavior related to linguistic analysis carries philosophical significance. A private language is not possible because the grammar of its words cannot be established. They do not belong to any communicative nexus of any form of life.

Wittgenstein's discussion of psychology is convolved and difficult to grasp. There is a lot of discussion of "pain" and who can ascribe pain to whom. Psychologically this has created a lot of discussion. And the possibility and scope of a private language is the topic of a huge body of investigations. In this work, however, we are chiefly concerned with Wittgenstein's relevance for the natural sciences. Thus his analysis of meaning turns out to be important, but the relevance of his discussion of psychology is marginal. Consequently only the main features have been summarized, however central the material may be to Wittgenstein's thinking.

5.4. What can we really know?

5.4.1. Moore on knowledge

In two lectures [96] and [97], Russell's colleague George Edward Moore argued for a commonsense view of knowledge we can and may be in possession of.

In [96] Moore lists a series of statements which must be accepted as true by anyone:

> There exists at present a living human body, which is *my* body. This body was born at a certain time in the past and has existed continuously ever since.
>
> Ever since it was born, it has been in contact with, or not very far from the surface of the earth.
>
> There has also existed many other things having size and shape in three dimensions, from which it has been at *various distances.*

According to Moore, there are also other human beings:

> What I assert is only that each of *us* has frequently *known everything I was claiming to know about myself and my body.*

The rest of Moore's paper is devoted to formulating these statements and delimiting his philosophical convictions from the attitudes of other philosophers. Moore claims that common sense has to accept the truisms that I exist as well as others exist and live in an ambience of a variety of things.

Kant at one point expresses his difficulties in proving the independent existence of external things. Responding to this argument, Moore argues in [96] for the existence of objects outside consciousness having its subjective experiences:

> I can prove now, for instance, that two human hands exist. How? By holding up my two hands saying, as I make a certain gesture with the right, "Here is one hand." and adding as I make a certain gesture with the left, "and here is another." And if, by doing this, I have proved *eo ipso*, the existence of external things, you will all see that I can also do it now in numbers of other ways; there is no need to multiply examples.

The rest of Moore's paper is devoted to making his position sharper and convincing the reader that his proof really constitutes a valid demonstration.

5.4.2. Wittgenstein on certainty

Wittgenstein was acutely aware of the central importance of Moore's problem. He argues:

"I know that I have two hands." Why not say: "I cannot doubt that I have two hands."

"I know" is the same as "It makes no sense to doubt."

Wittgenstein worked industriously to formulate his own vision of what needs and can be stated with conviction. The resulting parts of the *Nachlass* record the very last days of his life. The final entry is dated April 27, 1951, i.e. two days before his death.

The notes are published in 1969 [98].

Doubt or knowledge?

What we can ask is whether it makes sense to doubt. "I know" is in place when there is no doubt. Hence even when the expression of doubt is unintelligible, some certainty is more or less satisfied. A doubt that doubts everything would not be a doubt. If you tried to doubt everything, you would not get as far as doubting anything. The game of doubting itself presupposes certainty. Responding to Moore, Wittgenstein retorts: "If I want to doubt whether this is my hand, I avoid doubting whether the word 'hand' has got any meaning. I seem to know something anyway."

Whether a proposition is claimed true or false depends on what I may count as determinants of that proposition. Knowing means one has proper grounds for one's statement. What counts as an adequate test belongs to logic. It belongs to the description of the language game. The truth of empirical propositions belongs to the frame of reference. If you are not certain of any facts, you cannot be certain of the meaning of your words either. Can it be seen from a rule, what circumstances logically exclude a mistake in the employment of the rules of calculation?

Within the language game, this implies the possibility to imagine how this knowledge is argued. When the language game changes, there is a change in its concepts and with the concepts, the meaning of words changes. Giving grounds must come to an end sometime. The end is not an ungrounded proposition, it is an ungrounded way of acting. It is our acting which lies at the bottom of the language game. It belongs to the scaffolding of our thoughts.

Wittgenstein realizes that doubting can only work if there are sufficient grounds to support the doubts which are put forward. For more arguments, see the sections on certainty below.

The received view

We learn the nature of calculating by learning to calculate. One does not get one's picture of the world by satisfying oneself of its correctness. It is the critical background against which one judges what is true or false. The game can be learned practically without learning the rules. This is how one learns the game of "judging". Experience is not the ground for promoting our game of judging. Nor is its outstanding success.

Doubt is not necessary even when it is possible. Wittgenstein wants to regard man as an animal to which we grant instinct but not ratiocination. A creature in a primitive state. Any logic good enough for primitive means of communication needs no apology. Are we to say that the knowledge that there are physical objects comes very early or very late? "There are physical objects" is a logical instruction about the object.

We do not learn the practice of making empirical judgement by learning rules. A totality of judgement is made plausible to us. When we begin to believe anything, what we believe is not a single proposition but a whole system of propositions. It is not a single axiom that strikes one as obvious, but a system where consequences and premises give one another mutual support. What we hold fast to is not one proposition but a whole net of propositions. Our talk gets its meaning from the rest of our proceedings. If someone says he knows something, it must be that, by general consent, he is in a position to know.

As children we learn facts. The child learns by believing the adult. One learns an enormous amount and accepts it on authority. One must accept certain authorities to make any judgement at all. One finds some things confirmed and some disconfirmed by one's own experience. Doubt comes after belief. There are certain doubts a reasonable man cannot have.

This is how we learn science. Learning is based on believing. The system is acquired by observation and instruction. What grounds do we have for trusting textbooks on experimental physics? Should we say: "I believe in physics," or: "I know that physics is true"? Do we have grounds for letting ourselves be guided by the propositions of physics?

There is no sharp boundary between a methodological proposition and a proposition within a method. The lack of sharpness is between a rule and an empirical proposition.

The propositions of a world picture may be a kind of mythology, and a mythology may change. The same propositions may be treated at one time as something to test empirically or at another as a rule of testing. The physical game is just as certain as the arithmetic one.

To make an experiment you have to believe the apparatus exists. It belongs to our scientific investigations that certain things are indeed not doubted. Some facts are like hinges on which they turn. If we want the door to turn, the hinges must stay put. This is to be regarded as a *form of life*. Are a lot of our statements incapable of being wrong? The game proves its worth. This may be the cause of its being played, but it is not its ground.

The language games we participate in are learned without necessarily knowing the detailed rules. We accept the total framework not its independent bits. These may not be too unambiguous at all; the play may change when being played. We learn by accepting authorities. The natural stage of trust comes before the state of doubt.

5.4.3. von Wright on certainty

As usual, Wittgenstein's argumentation is multifarious, convolved and slightly incoherent. Thus his position vis-à-vis Moore's argument. His search for solid foundations had not produced a convincing collection of truths. He assures us that philosophy does not aim to change anything; language, mathematics, and science are left as they are. The role of philosophy is to observe and criticize the use and misuse of the formal systems. But except for purifying the common usage of the discourse, there is very little left to found the phenomena observed.

The situation is greatly clarified by von Wright [99]. His presentation is coherent and clear, and states very succinctly what can be concluded from Wittgenstein's writings. He expresses Wittgenstein's argument as follows:

In a genuine knowledge situation there must be grounds for knowing.

But the claim of grounds (evidence) has an end, a point beyond which no further grounds can be given.

All testing, all confirmation and disconfirmation of a hypothesis takes place already within a system.

In every situation where a claim to knowledge is being established or a doubt settled, or an item of linguistic communication is being understood, a bulk of propositions already stand fast, are taken for granted.

The system is not so much a point of departure as the element where the arguments have their life.

The propositions belonging to the system of pre-knowledge cannot be "enumerated" and laid down once and for all.

That which cannot be conceivably doubted cannot be known for certain either.

The system of propositions which constitute a world-picture not only has no fixed boundaries, it also has very inhomogeneous composition.

These subsystems are related to what Wittgenstein calls *language games.*

6

SHAKY FOUNDATIONS

6.1. In logic we trust?

6.1.1. Quest for certainty

We humans live in an insecure ambience. No one knows what ills tomorrow brings. Thus the role of religion and philosophy has always been to provide some security, some fixed points around which we may turn the events of our lives. Because the comfort offered by religion is based on blind trust, it has fallen to the lot of philosophy to provide logical arguments for the validity of, at least some, eternally necessary truths.

The irony is that philosophy has mostly served to pour cold water on any attempt to provide solid answers to questions relating to existence. Already Plato worked hard to abolish cheap metaphysical arguments based on clever linguistic manipulations. Plato's method was a penetrating analysis of the ability and limitation of the use of words. They can only carry meaning within the linguistic network, but to catch their meaning one has to locate them in the social context of human experience. This has ever since been the method of philosophy; clarity has emerged but no trust in eternal convictions has been assured.

Descartes started from the opposite pole of human life, the inner experience of one's own existence. Knowing a game is not a state of consciousness. What would it mean to state: "I am not conscious"? Can one pretend to be conscious? To claim consciousness is a proposition where no doubt is possible. If there are no unconscious thoughts, there are no conscious ones either. Being conscious of one's consciousness is taken to be a given datum, which can provide the rock on which to build the temple of reality. Unfortunately the foundation cannot be solid, the argument is blatantly circular, and only divine intervention can guarantee the solidity of the edifice. Then we are back where we started; only faith can provide faith.

The progress of science in the seventeenth and eighteenth centuries seemed to suggest that the facts of nature are necessarily true, the laws of science must be unequivocally necessary. However, Hume could counter this; no phenomenon can be repeated enough times to ensure that it will recur. However many times an apple is seen to fall to earth, there is no logical certainty that it will not, one day, fall upwards. Causality has to remain a hypothesis, ever to be threatened by falsification.

The latest effort to build a cathedral of human knowledge was initiated by the positivistic movement started at the dawn of the twentieth century. Accepting our sensory impressions as genuine signals from an existing reality, its adepts tried to construct a logically complete world for us to live in. Unfortunately human reality turned out to be vastly more multifarious than the positivists expected. Presently, the program of constructing a unique logical image of reality seems to have been abandoned as hopeless. This does not, however, mean that the effort has been useless. Along the way many small shrines have been erected instead of the hoped-for cathedral.

6.1.2. The frail temple of mathematics

The final desperate search for at least some certainty turned to mathematics. The statement "2+3=5" seems to be true beyond even unreasonable doubt. Geometry with its "truths" played a central role already in the thinking of the ancient Greeks. The validity of Euclid's demonstrations were long regarded as so certain that Kant found them to be necessary facts about a reality in no need of empirical confirmation, synthetic truth *a priori*. The emergence of non-Euclidean geometry put an end to that.

Then philosophy turned to algebra; the existence and algebra of integers were beyond doubt; according to Kronecker they were made by God himself. Independent efforts by Gottlob Frege in Germany and Bertrand Russell in England, aimed at creating a bedrock foundation for the by then impressive edifice of mathematical truth. Their activity was based on set theory and logic, concepts which they regarded as solid enough to be beyond doubt.

Frege considered mathematical objects to have real existence, and identified concepts with the set of elements falling under it; concepts have sense and extension (*Sinn* and *Bedeutung*). The universal application of this approach was frustrated by Russel's set of all sets not being members of themselves. Further paradoxes have been constructed, showing that sets are as dubious as the concepts they are supposed to justify. The program to found mathematics has never quite recovered, but some new directions have emerged.

Intuitionism

In Holland, the mathematician L. Egbertus J. Brouwer adopted an extremely critical attitude to mathematics. The basis is to abandon the principle of the "excluded middle", which will make all indirect proofs invalid. Mathematical existence can only be shown by constructing the object claimed to be. Also simple propositions are not considered to have truth values until this has been explicitly proved. The statement "At some position in the decimal expansion of π there occurs 100 consecutive zeroes" is not true but also not false until this extremely unlikely sequence has been found. All mathematics has to be constructed on direct proofs and explicit constructions.

This movement basing the effort on an ever-growing but eternally unreachable goal has been termed "intuitionistic". Much constructive re-creation of known mathematics has been possible, but central results of classical analysis have turned out to be beyond reach. Active mathematicians regard the framework as too restrictive, leading to much too poor a mathematical system.

As a philosophical approach to mathematics, intuitionism has opened up new views: what can be made certain beyond doubt? It is maintained that Wittgenstein listened to a lecture by Brouwer in Vienna in March 1928. This was so challenging that it caused Wittgenstein to return to philosophy. Thus he too was motivated by the quest for some foundation for the conception of reality. As we will see, he too ended up causing more uncertainty affecting human life.

6.1.3. Interpretation of mathematics

In Göttingen, David Hilbert was unquestionably the leading mathematician at the beginning of the twentieth century.

Hilbert's young friend was Leonard Nelson, a philosopher. He had a degree from Berlin, and tried to get the Habilitation from Göttingen. He came in conflict with Husserl and his thesis was rejected by the Faculty. Hilbert supported him and after a few years he got the Habilitation accepted.

When Russell and Whitehead published their *Principia Mathematica*, Hilbert started to support Nelson more. His interests were the logical foundations of philosophy and principles of ethics. Opposed by Husserl, Nelson became an associate professor only after the war.

In 1918, Brouwer questioned the foundations of mathematics (in three papers less than seventeen pages). He was well established; in 1911 he had attracted attention by solving a topological problem. Hermann Weyl joined the critical movement with his essay on the introduction of the continuum [72]. Brouwer was to become the main father of the intutionistic school of mathematics [73].

At the Third International Congress of Mathematics in Heidelberg in 1914, Hilbert announced that the foundations of mathematics had to be investigated. This is the first time that proof itself was to be the subject of an investigation.

Hilbert returned to the foundations of mathematics in 1917. He was joined by the young Paul Barnays who had an interest in philosophy and had been in contact with Nelson. They met in Zürich, and Hilbert arranged for Barnays to come to Göttingen.

In the summer of 1920 Hilbert concentrated on the possible axiomatization of atom mechanics. After this endeavor had been without success, he continued the effort to formalize the foundations of mathematics,

In 1922 he was upset by the progress of Brouwer's intuituism and set out on his formalization program of proofs thus trying to save classical mathematics through finitistic methods.

As noted above, Wittgenstein took up philosophy again when he encountered the intuitionistic mathematics of Brouwer.

At the Königsberg conference 1930, Waismann presented some of Wittgenstein's results. This was, however, overshadowed by the announcement of the incompleteness result. Hilbert's program too was hampered because of Gödel proving the incompleteness of formal systems. Hilbert, however, continued to work with foundations. In 1939 his assistant was Gerhard Karl Erich Genzen.

Genzen managed to introduce more powerful proof methods, thus contributing to the new emerging field of *metamathematics*. Genzen died in 1945 after imprisonment in Prague.

6.1.4. The beauty of formalism

David Hilbert had initiated the most abstract approach to the totality of mathematical relations. In 1899 he formalized the foundations of geometry; a work which achieved immense popularity. Only later did he embark on the grand endeavor to master all of mathematics. In explicit opposition to the intuitionist activity, he attempted to formalize the very process of derivations. For this he needed, in addition to the axioms, the transformation rules which make one valid statement imply another one inheriting the truth from the axioms. Thus the whole structure becomes a game with symbols, where all valid theorems can be shown to follow quite formally from the assumed axioms. This is the origin of *Proof theory*, which is still a vivid and developing discipline of logic.

To illustrate the process I introduce a toy world of formals:

Objects: $\{x_1, x_2, x_3\}$
(syntax)

x_i is an undefined element

Given (Axiom): $\{a, b, c\}$

Transformation rule: $\mathcal{T}\{x_1, x_2, x_3\} \Rightarrow \{x_2, x_3, x_1\}$

From the axiom and the transformation rule we obtain the objects

$$\{a, b, c\}, \{b, c, a\}, \{c, a, b\}$$

If these objects are labels for theorems, we have derived two new ones from the axiom. The formal manipulations, however, need no interpretations, all we need to know is their transformation rules.

In the formalization of mathematics we need to consider some aspects of the system posited: The one is *completeness*. We can add no element to the system without introducing a contradiction (*vide infra*). The system above is clearly not complete. We may add the transformation rule

Transformation rule #2: $\mathcal{K}\{x_1, x_2, x_3\} \Rightarrow \{x_2, x_1, x_3\}$;

now we may consistently derive the validity of the additional objects

$$\{b, a, c, \}, \{a, c, b\} \{c, b, a\}.$$

This makes all allowed objects derivable, which makes the system rather dull but it is a transparent example of a formal deduction system.

The more interesting aspect is the *consistency* of any formal system. This demands that we cannot derive any statement (A) together with its negation ($\neg A$). This is not the case in our model but the criterion is essential enough to deserve a closer investigation.

Formal approach: We introduce a set of statements: $A, B, C, \ldots$. These take the values true (T) and false (F). For these we claim to have derived the following facts:

$$\text{(I)} \quad ; \quad A = T$$

$$\text{(II)} \quad : \quad \neg A = T \quad \Rightarrow \quad A = F$$

We perform the following manipulations

$$A = T \Rightarrow A \vee B = T$$

Combining this with (II) we get

$$(A \vee B \,\&\neg A) \Rightarrow B.$$

Thus in a system with contradiction any arbitrary conclusion (B) is derivable.

This argument looks like a trick valid only inside the formal structure. It is worth our while to see how it may operate in a real life situation:

- Adam calls his neighbor Niels telling that the common water well is not working. Niels asks: "Are you calling from home?" Adam: "Yes." (*Statement 1*).
- The serviceman arrives and sees Niels in the garden. Serviceman: "Can I get into Adam's house?" Niels: "Yes he or his wife Eve is at home." (*Based on Statement 1).*
- Serviceman sees Adam in the pub across the street. Adam can thus not be at home (*Statement 2*).
- Serviceman concludes that Eve is at home. This is a correct conclusion, but it need not be true.

The contradictory *Statements 1 and 2* induce a conclusion that may, in fact, be wrong. This is a common situation when a formal system contains contradictions. Consistency of a system is thus proved if one can show that there exists a statement which is unprovable.

Gödel incompleteness

Hilbert's dream was to find a machine capable of investigating any well-formed statement and tell if it was derivable as a theorem or not. This was shattered when Gödel proved his incompleteness result. This theorem was considered above, but its central role in mathematics and logic justifies that we look closer at its structure:

Set the statement A to be

$$A = \text{“this statement cannot be derived”}.$$

Consider derivations in a *formal system* **F** where A can be expressed. Let us start listing all theorems derivable consecutively in the sequence

$$X_1 \Rightarrow X_2 \Rightarrow X_3 \Rightarrow \dots$$

Assumption **X**: For some n, $X_n = A$.

Thus we have derived A, from assumption **X.** But A claims that this is not possible; thus we have a contradiction. **X** can consequently not be true: $\neg$**X** must hold. A cannot be derived, which is just what it states: It is consequently true but underivable. Truth transcends the concept of derivability.

We can, of course add A to the original axioms of the system, but then we can find another statement which is not derivable. Some conclusions are, however, possible:

The formal system used by Gödel is the algebra of numbers. The theorem is dependent on the assumption that their algebra is consistent, which cannot be proved within Gödel's logic. The result has, however, been obtained by methods beyond those used by Gödel, but only at the expense of additional dubious assumptions. However, his procedure itself suggests that algebraic consistency is the case. Because we have extracted a theorem which is not provable, we may conclude that the structure utilized contains no contradiction, as was explained above.

The presentation above looks at the formal structure as distributing derivability along a chain of sentences. This approach reminds us of a computer program. Indeed Turing managed to obtain the Gödel result as a criterion for *computability.* Logically equivalent results were obtained by Church, hence their result is referred to as the Church-Turing thesis.

An additional conclusion relates to the application of formal mathematical science. Consider all sentences, formed correctly according to the syntax of the system. If their validity would uniquely be determined from within the formalism, we could not use them in science. We would encounter Wittgenstein's world where all true statements are logical tautologies. There must be well-formed statements in the system whose validity is not a consequence of the logical structure. These contingent statements are the ones we may test by an empirical method thus obtaining knowledge about the structure of reality. Gödel's theorem seems to lie behind the utility of mathematics in science. Even if our structures are consistent as they stand it is, of course, beyond our capacity to overview the whole edifice of natural science.

Löwenheim-Skolem ambiguity

By definition, formal systems are "formal". They cannot convey facts outside themselves. It is thus a doubtful enterprise to graft an abstract structure on the material derived from empirical actions. In spite of this, the mathematical structuring of physical data has proved unexpectedly successful. All the time, however, one has to keep in mind that parts of our theoretical models, our maps of reality, will derive from the method of mapping without direct relevance for the structure of reality. Many kinds of maps can be used to describe the same terrain, and only the use can decide their relative merits. In particular, the mapping can be used to announce what not to expect from a theory.

The theoretical description of empirical reality is based on *models*. In physics, this means finding a formal mathematical structure which can be used as a framework for ordering the numerical results of observations. When such a frame encompasses a wide range of phenomena, we claim to have discovered a fundamental theory, a *Law of Nature*. This holds with respect to Newtonian mechanics, Maxwellian electrodynamics, and the more ephemeral theories of modern physics. It remains, however, prudent to remember that these laws exist primarily as our mental constructions and their absolute mode of being remains an enigma.

In philosophy the term "model" has got a slightly different meaning. It denotes a realization of the axiomatic structure introduced and investigated as a formal axiomatic system. Axioms and rules of derivations do not tell what the basic elements are, they are taken for granted. A model is the devising of such an interpretation. Euclidean geometry of lines and points is a model of Hilbert's axiomatic system for "geometry". In physics, the representation of a group is an example which is commonly used and well understood. It is also well understood that the formal structure admits several different realizations as representations. Thus the multitude of elementary particles can all be identified as different representations of the Lorentz group.

The ambiguity of models (representations) has its counterparts in the world of logic. This is based on theorems introduced by Skolem and Löwenheim. These results are derived for the logical framework of formal systems. The heuristic consequences include a statement that if a logic system has a valid model at all, it has a model at most denumerable. This has grave consequences for the practical applicability of this structure, because reality clearly does not contain only denumerable numbers of entities. Thus it follows that, if such a model is embedded into a broader category, the ensuing model cannot be uniquely defined by the denumerable original information. As all empirical measurements may satisfactorily be represented by a denumerable data set, the application of any conclusion must be compatible with a multitude of extended models. No empirical material can definitely verify a theory, it can only, at its best, falsify one.

How is it then possible that our theories can describe empirical reality which we model by continuum structures? The answer seems to lie in the very concept of being "denumerable". This depends on the elements included in a set: the set of rationals is denumerable, the set of reals is not. The conclusion also depends on the way of defining the method of comparison. We assume that we have sets $A = \{a_i\}$ and $B = \{b_j\}$. These have the same "cardinality" if we find a one-to-one mapping between the sets

$$\Phi(a_i) \Longleftrightarrow b_i; \quad \forall i.$$

In a formal structure, the validity of this statement will depend on the store of possible mappings Φ we have at our disposal. The size of a set is thus not unambiguously fixed by its elements alone. Sets of differing cardinality may, in principle, become equinumerous if we allow a different set of mappings.

The situation may be elucidated by the following (trivial) example: Let the original set be given by

$$X = \{n_1, n_2, n_3 \,|n_i \text{ is an integer}\}.$$

Define the set of mappings $\{\Phi_q\}$ by setting

$$\Phi_q X = \{q^{n_1}, q^{n_2}, q^{n_3}\}.$$

Now for

$$q = \exp(2\pi i),$$

there is no one-to-one mapping between X and $\Phi_q X$; hence they do not have the same size. However, if we choose $q = \pi$, we get equinumerous sets. The example is indeed trivial, but it serves to demonstrate that the size, i.e. cardinality, of a set depends on the mappings we have available for a comparison of the Cantor type.

Thus the application of formalism to empirical data does not single out a unique theoretical model. On the other hand, the Löwenheim-Skolem denumerable model may be applied to continuous data by a reorganization of the range of possible mappings.

Thus theories may work, albeit only heuristically, but even the most successful theory cannot condition reality in an absolute way. Logic has its limits and human knowledge will always remain restricted.

6.2. Mathematics as a sovereign world

Wittgenstein's *Nachlass* contains much material relating to his two published works. The same questions are taken up and his views are changing and developing. In addition, the terminology of the translations is not always consistent or even appropriate. It is not the purpose of the present writing to extract a unified and consistent presentation of the totality of his work. Here I only choose to summarize the material without detailed references to the published texts. Detailed listing of the sources would make an unwieldy presentation. In addition, many points are taken up in several places. Wittgenstein's terminology is varying, but the issues at stake tend to remain fairly persistent. A living being communicates only with formal signs, so they must be something essential. But what is a living being? In the text below, I try to extract statements in which the positions of Wittgenstein's later views are illustrated. I do not try to interpret his approach in a consistent and clear manner. Such an integrated view is neither possible nor was it probably intended by Wittgenstein.

6.2.1. Logic and language

Wittgenstein considers logic and mathematics as one stage of human acting. Both are parts of the structure of communication consisting of the autonomous and intertwined language games. Like Bohr, he realizes that the use and misuse of language is necessary to judge the possibility and role of language:

Wir kämpfen mit der Sprache.[1]

Human beings are entangled all unknowing in the net of language. The role of the acting individual is stressed by Wittgenstein. Even if there is a world of facts, their interpretation has to be constituted by the individual. He says [91]:

Philosophie dürften man eigentlich nur dichten.

The English translation of this is: "Philosophy ought really to be written only as a *poetic composition.*" The translation of *dichten* seems to me mistaken. I interpret this as:

Philosophical writing really ought to be fiction.

This is supported by Goethe's autobiographical work *Wahrheit und Dichtung* which is usually translated as "Truth and Fiction". This was published already in 1833.

6.2.2. Language

Often it is claimed that the later Wittgenstein differs from the early one. This is only partially true, but some aspects are explicitly brought up by himself.

Elementary propositions are not independent as claimed in the *Tractatus* [10] In fact, in a system the assignment of a property implies the denial of other properties of the system. Logical attributes have inner connections. To consider a "fact" as a "complex of objects" is misleading [82]. Logical products are not hidden in sentences

It is wrong to think that analysis is to bring out what is hidden. What is "hidden" depends on the method of discovery. Elementary propositions are conditioned by each other.

Expecting identity in form between thought and reality is an error. However, thought must have the same logical form as reality. An object cannot have two colors. This is the negation of an a priori proposition; hence it is *a priori* itself. In quantum physics, this may be compared to the complementary properties.

Wittgenstein is indifferent to solutions of purely scientific problems. He is not concerned with facts of physics except in so far as they determine the laws governing how things appear. Physics strives to produce "truth" but our descriptions require sense.

[1] We are struggling with language.

Compare the giving of information by maps and drawings with the giving of information by sentences. The sentences are no more arbitrary than the drawings are; only the words are arbitrary. On the other hand the projection of the map is arbitrarily devised by us.

If we want to study problems of truth and falsehood, agreement and disagreement of propositions with reality, the nature of assertions, assumptions, and questions, we should look at language, without the confusing background of any complicated processes of thought.

In every serious philosophical question, uncertainty extends to the very roots of the problem. The rule-governed nature of our language permeates our lives.

Language contains the same traps for everyone as long as there are words like: "to be", "identical", "true", "false", "possible", as well as "flow of time".

The terms used in discourse have the haziness of concept-words in our language.

We cannot achieve greater generality in philosophy than in life and science. When we talk about language one must use the language of practice. How strange that we can do with the one we have.

A proposition is conceived as a picture of the state of affairs; it describes how things stand if it is true. Thus it shows the possibility of the asserted state. It can set forth what is not the case. It depends on our grammar what is possible and what is not, i.e. what grammar permits. "Possible" means conceivable. But possibility does not yet guarantee existence. How strange that we should be able to say that a state of affairs is inconceivable. Looking somehow presupposes that I know what I am looking for, without what I am looking for having to exist.

What expectation has to do with reality is that it refers to another point in the same space. You can only search in a space, because only in a space do you stand in relation to where you are not. How do we describe that which does not exist?

All forms of speech are taken from the ordinary physical language and cannot be used in epistemology or phenomenology without casting a distorting light on their objects. To say that a straight line can be drawn is to say that it makes sense to talk about this line.

Anything can picture anything. Every projection must have something in common with what is projected. Wittgenstein uses the word "projection" for what physics terms "correspondence rule".

Language cannot express what belongs to the essence of the world, it can only say what we also could imagine differently. The self-evidence of the world is expressed in the very fact that language refers only to itself and can mean only it. Wittgenstein uses early the term "systems of communication": Does this give "language games"?

Grammar is not accountable to any reality. It is grammatical rules that constitute meaning, and so they are not themselves answerable to any meaning and to that extent they are arbitrary. The rules of grammar are arbitrary in the same sense as the choice of unit of measurements. Mathematical texts are not descriptions of "something"; they are the thing itself. Mathematics consists only in algebraic operations and derives no meaning from its outside. We cannot describe mathematics; we can only do it.

To invent a language may be to invent an instrument for a particular purpose on the basis of the laws of nature. The concept of language is conditioned on the concept of communication.

Laws of logic are arbitrary. We can have different rules of "all", "some", "not", and "any". They have different meaning if applied to cardinals or reals. Our world looks quite different if we surround it with different possibilities. The difficulty in philosophy is to find one's way about. There is no definite class of features characterizing all uses of the word "can".

Wittgenstein introduces the central concept of "logical multiplicity": The negation of a proposition describes the same situation but with different polarity. When a negation is meaningful, we have to know what is negated.

A negative proposition has the same multiplicity as its positive counterpart. According to Wittgenstein, they make just as much sense. To know what is the case, we need to know also what is implied not to be the case. The superfluous part of the language, Wittgenstein terms "wheels turning idly".

A symbolic system with the right multiplicity will render syntax superfluous. On the other hand, the syntax may render symbols superfluous. An incomplete symbolic system can always be amended by adding rules of syntax. Only physics can tell how many concepts are needed.

The multiplicity corresponds not to one verification but to the law obeyed by verification. Two propositions must have the same sense if every possible experience confirming the one also confirms the other. Your knowledge about the general fact has the same multiplicity as your reason. In order to guide one's action, the negation must have the same multiplicity as the proposition it denies.

The grammatical rule for the terms of a general proposition must contain the multiplicity of possible causes provided for by the proposition.

The best comparison for every hypothesis is a body in relation to a systematic series of views from different angles.

It is always single faces of a hypothesis that are verified. If that face is laid against reality the hypothesis becomes a proposition.

A command must have the right logical multiplicity, i.e. the same as its actions. Controls must have the multiplicity of what the machine is capable of making. To expect more is to talk nonsense in language. Expectations and the verifications have to have the same multiplicity.

Epistemology pays no attention to the truth or falsity of contingent propositions. Epistemology highlights everything that can be thought.

6.2.3. On objects

By correlating the names to facts, one knows how many entities there are. If I write "AA" on the board, have I written one letter or two? Depending on whether "A" means this shape or the character.

If "objects" are elements of representation, there cannot be ∞ many of them. The formula for real numbers is not to say that it holds for *all* real numbers, but that given an arbitrary real number, it holds of this. I can say "Wipe the table!" but not "Wipe all the points of the table!"

Points cannot be seen and singled out. Euclid never stated that lines consisted of points; they were logically independent entities. Two lines define a point, but no property defines all points. The point of intersection of two lines is not a common member of two classes of points, it is the meeting of two laws.

According to Wittgenstein, the integers can be used to transfer a property through induction, which can be represented formally as

$$[f(1); f(\xi) \Rightarrow f(\xi + 1)] . \tag{8}$$

This does not, however, justify

$$\forall x : f(x). \tag{9}$$

Wittgenstein says that this assumption is one of his early mistakes eliminated later. The statement

$$f(a) \& f(b) \& f(c)$$

is, however, equivalent with

$$f(x); x \in \{a, b, c\} \equiv A. \tag{10}$$

The form (10) allows that there are other elements x in addition to A, the form (9) does not. For the former it makes sense to add

$$\exists x, x \notin A,$$

but if A as a set is complete, it tells everything. This opens the question of the conjugate set, is it possibly empty? Only if there exists a set $E \neq \varnothing \equiv \{\}$and $A \in E$, can we define a complement

$$\widetilde{A} \equiv E \ominus A.$$

If we list A we do not need to say that "there are no others". If $\widetilde{A}$ is nonempty, the set

$$P = \{p \mid \neg p \in A\}$$

has got "extension", otherwise not.

This relates to the meaning of negation. Namely, stating P, we can give meaning to $\neg P$ only if we may construct a set

$$\{x \mid P(x) = \text{false}\} .$$

It is not sufficient to state

$$\neg P(x),$$

i.e. x does not imply that P is not true. We have to verify that P is false.

The meaning of a proposition is its mode of verification. The meaning of a word is its use.

The subject constructs his formal space but the result is objective. In physics space is Space. Two elementary propositions cannot contradict each other. A classical point mass can have only one velocity at a time. There can be only one charge at a point in an electric field, only one temperature on a surface, the body can have only one temperature. No one can doubt that these are self-evident and their denial is contradictory.

It is not a proposition we put against reality as a meter stick, it is a system of propositions.

6.2.4. Signs and symbols

Wittgenstein introduces "shadow of a fact" = "proposition and its sense".

He also introduces the concepts of "sign", which is only a (simple) figure, a scratch or sound, and "symbol", which carries a meaning. This attributes an "intention" when it is used. The significance of a sign can only derive from its participation in a symbol. It is significant only within a system. The genuine criterion for a formal structure is precisely which propositions make sense for it—not which are true.

The place of a symbol in language is given by its use. A map is a picture of reality only through its intention. Explanation or experience can only add to the meaning of the symbol. The word says nothing about the world; a proposition says something and thus can be false. We understand a symbol only as part of a system described by its grammar. The symbol is a sign together with all the conditions necessary to give it significance. Understanding the symbol does not tell if it is true or false, for this you need the fact too. There are no necessary facts. Everything that can be described must be capable of being symbolized. Grammar gives the rules for the combination of symbols.

Imagine someone who knows his way around the city perfectly, i.e. finds the shortest way, but is unable to draw a map of the city. Understanding is effected by explanation but also by training. Knowledge by words is not a translation of something that was there before. What is the language game of "to know"?

Does a border case mean that the word "plant", in all cases, is infected by uncertainty? Would an exact definition clear up the ordinary use?

The representation ought only to contain adequate concepts. If they are not needed for the use of the conceptual system, they are Wittgenstein's "wheels turning idly".

If there are several languages; what they have in common is then what they depict. What is invariant under change of representation is a fact. This suggests the "invariants" of Eino Kaila and Felix Klein.

Wittgenstein also discusses "external" and "internal" relations. Internal relations derive from the grammar of a concept, whereas external ones relate to pairs of objects (or larger collections). A concept is not essentially a predicate. An internal relation lies in the character of things. It is never a relation between two objects but a relation between two concepts. The electron's internal relations are correlations of charge, spin, and rest mass. Their interactions are external.

The method of answering a question tells what the question is about. Where you can ask a question, look for an answer; where you cannot look for an answer, you cannot ask either. Nor can you find an answer. Only where there is a method of solution is there a question. Do not ask for a definition, get clear about the grammar. To construct a language that covers all situations is a conceptual absurdity.

A philosophical problem is of the form: "I do not know my way out." Such a philosophical problem must be completely solvable. "Realism" and "Idealism" are metaphysical terms; they try to say something about the essence of the world.

The solution to the riddle of life may not make life easier. It was possible to live even before it was recognized. If there were a result of logic, even before this was discovered it was possible to think.

6.2.5. Procedures of mathematics

Wittgenstein tries to show that a mathematical discovery is rather a mathematical invention. Solution of a mathematical problem never helps in philosophy.

Russel's calculus is a bit of mathematics. It is possible as operations even without the formulation in terms of classes. Philosophy may not interfere with the actual use of any language; it can only describe it. It cannot give it any foundations either, A "leading problem of mathematical logic" is a problem of mathematics like any other.

Tables, ostensive definitions and similar instruments we may call "rules". The use of a rule is explained by a further rule. In mathematics everything is algoritm and nothing is meaning. Set theory preserves the picture of something discrete but fails to carry the argument to the finish:

Number is an attribute of a function defining a class, not a property of the extension.

Fitting the number of primes to a logarithm is like an experiment in physics. We must think of "number" as we think of "length", or "weight" and of counting or correlating as we think of weighing or measuring. We may say that the body does not have any weight at all except when being registered, or no definite weight except when being weighted.

By the means of our intellect we look into a certain realm and see that the propositions of logic are true. This makes logic the science of the intellectual realm. Number is an external property of a concept and an internal property of its extension (the list of objects that fall under it). Extension of a concept is a list.

Knowing how to go on must be based on *prior information.* Complete induction is not expressible as a proposition:

Let us define the scheme **A** by the operations:

$$\begin{aligned} \exists n_0 : \;& f(n_0) = \quad T \\ \& \;\;& f(n) \Rightarrow \; f(n+1)\,. \\ \Rightarrow \;& f(n) = T;\; \forall n \geq n_0 \end{aligned}$$

In words: the property tested for by $f(n)$ is true for all values of n. The scheme **A** is not a proposition because it cannot be negated; what would be the meaning of $\neg$**A**?

It is impossible to extend a system; many propositions in the extended system lack meaning in the original system. Thus it is not just a "completion". We know as much about our formalisms as God does. What we do not have, we do not have. There are surprises in reality but not in language. Language can expand. Is it an accident that in order to complete the sign system we have to go outside the written and spoken language? Wittgenstein is presumably thinking of extensions of the number system. The integer "2" is not the same entity as the real number denoted in the same way.

The actual infinity is a mere word. What can it mean to say that a straight line can be arbitrarily prolonged? Further: there is no such thing as an infinite list.

The idea of an infinite chain of reasons follows from a confusion like: A line of certain length consists of an infinite number of points because it is indefinitely divisible. But no algorithm can produce all its points, including every irrational one.

Proofs of equal cardinality by matching figures are like experiments. If the patterns do not match, we seem to have learned something about reality.

A contradiction is manifest only if encountered. Hidden contradictions are not to be worried about as long as the game can go on. A contradiction does no harm until it is made to operate. Language games are languages complete in themselves. We have not encountered contradictions in mathematics. A mathematical statement has no sense before being proved true or false within some system. What is not proved is not known.

A proof implies only what it proves. In mathematics different proofs prove different things. A proof of existence gives the meaning of "existence" in that theorem. Of course, we may not and cannot deny the essential existence of anything. Some philosophizing mathematicians are not aware of the many different usage of the word "proof". All our collected knowledge fails to prove God's existence or lack thereof.

Mathematics is a game. Since mathematics is a calculus, and hence is not really about anything, there is no metamathematics. Metamathematics is mathematics in disguise. One cannot compare two classes of consequences unless one forms a new system in which both classes occur.

To someone saying that he has discovered that between the rational points of a line there are more points, we reply that he has not discovered new points but he has constructed the new points. How do we know that a theorem proved by transfinite methods cannot be violated by a concrete example? This is the mathematical problem of consistency.

"Formalism" versus "contentful mathematics" is like idealism versus realism. It will become obsolete in that each party makes unjust assertions at variance with the day-to-day practice.

A calculus cannot decide a philosophical problem.

Logic and mathematics are not based on axioms any more than a group is based on the elements and the operations that define it.

Every mathematical calculation is an application of itself and only as such does it have sense. Natural numbers and others have forms adapted to reality of things. Complex numbers are given by visualizable manifolds.

The symbol of a class is the list. We may compare mathematical equations with significant properties not with tautologies. This makes intelligible the attempts of formalists to see mathematics as a game with signs.

A generalization cannot be both empirical and provable. You can not talk about all numbers, because there is no such thing as "all numbers".

Surely it is impossible that mathematics should depend on hypotheses of physical space

6.2.6. *Reasons and causes*

A reason is a reason only within the game. Only a hypothesis can derive from a cause.

A mathematical argument can have a reason. An experimental evidence can have a cause.

Why should what finite and transfinite numbers have in common be more interesting than what distinguishes them?

Present foundations of mathematics treat it as if it could be reduced to something else. It is not to reduce it but to treat it in a different way. Wittgenstein does not like the identity "$x = x$". Identity in logic is equivalent with having all properties the same. This is not possible in physics. For objects, properties are had or not had. We recognize the different "identity" of the sentences: "I did what he did" and "I did what he ordered".

We must distinguish between necessity in the system and necessity of the system.

Would it matter if a physicist used a calculus which subsequently would be found to be inconsistent by mathematicians? No!

6.2.7. *Wittgenstein on foundations of mathematics*

Wittgenstein worried about the role of mathematics and its foundations. Why is it necessary that $2 + 3 = 5$, indeed, why is it almost obvious? The extensive remarks are scattered in many books publishing the *Nachlass*. Only parts of the works are covered in a more systematic way, *vide infra*. A recent survey of Wittgenstein's view of mathematics is given in [100].

Worlds are created

To know how to count is to know how to continue. Counting and calculating is not simply a pastime. It is a technique which is employed daily in the most various operations in our lives. And that is why we learn to count as we do with endless practice and merciless exactitude. But is this counting only a habit? Is there not also a reality corresponding to this sequence? The truth is that counting has turned out to pay. It is usable and above all it is used.

Conjecture in mathematics is like a hypothesis in physics. Different symptoms can be symptoms of the same thing. Logic is a kind of ultra-physics. The description of the logical structure of the world is perceived through some kind of ultra-experience. It is as if a mathematical conclusion "were there" already before we state it. It looks as if the formation of our concepts guided our experience into particular channels so that one experience is now seen together with another one in a new way. We might speak of a kind of "Alchemy" in mathematics. Can we envisage how to think the unthinkable? The synthetic character of the propositions of mathematics appears most obviously in the unpredictable occurrences of prime numbers. Wittgenstein would like to be able to describe how mathematics appears to us as the natural history of the domain of numbers and concomitantly as a collection of rules. To be able to make an assertion, I must do so with reference to its sense not to its truth.

Mathematically the system is, so to speak, a world. The forms and the rules of syntax are equivalent. It is not enough to say that p is provable, but we must also say in which system p is expressible but not provable. If p is transferred from one system to another, it does in reality change its sense.

We cannot learn anything new about an object in mathematics. Mathematics cannot be incomplete. The completely analyzed mathematical proposition is its own proof.

Axioms are propositions of syntax. A consistency proof cannot be essential for the applications of axioms. The total system of algebraic propositions corresponds to a system of inductions.

Wittgenstein considers set theory to be wrong because it presupposes a symbolic system which does not exist.

Mathematicians create essences. They are not an essential property of an object but rather the symbol of a concept. Samples of numbers can be used to unfold their properties. We do not need to unfold physical properties; they are manifest. Internal properties persist always in the whole which they constitute independently of any external accidents. They are not subject to wind and weather like physical properties.

Thus, according to Wittgenstein, mathematicians create closed worlds which display essential properties not intentionally imposed on the system. The systems investigated may be suggested by empirical experiences but must not be restricted by these. To uncover mathematical properties we use proofs.

To prove or not to prove?

A proof is accepted as a paradigm of a pattern derived by rules and applied to patterns constructed from empirical propositions. One writes down certain signs guided by certain paradigms. No sign should be overlooked and what one gets is what follows. This also follows when we look through a completed proof. Perspicuity is to be part of the proof. One may know a proof thoroughly and not understand what has been proved. One may form a proposition grammatically correctly without understanding its meaning.

Rules belong to the beginning, a preparation. For the construction of the system in which the law lives out its life.

A proof as an "experiment" results in something larger than the outcome of the proof. The proof shows what ought to come out. The result of a calculation is the final proposition, the result of the experiment is how one is led to the proposition. It must be easy to write down exactly this proof again. A proof ought to show not only that this is how it is, but this is how it has to be. Consider the law for the occurrence of the digit "*7*" in π; we do not know this law. This also depends on our decimal system. A number is the result of an arithmetic experiment so the experiment as a description is absurd.

When mathematics is divested of all content, it would remain that certain signs can be constructed from others according to established rules. So much is true about saying that mathematics is logic; it moves within the rules of language. This gives it its peculiar solidity, its unassailable position. It even forms new rules: a mathematician is an inventor not a discoverer. The application of the calculus must take care of itself.

If mathematics is a game, in proving we need never appeal to the meaning of the signs, i.e. to their extra-mathematical applications. But what does appealing to it mean? Does it mean leaving mathematics and returning to it again? If the law of excluded middle is ungainsayable, such an understanding is somehow wrong.

Demonstration of e.g. addition by sets of objects works only if these objects are known to retain their individuality. The number of objects can be utilized only if there are objects which retain their number. What is infinite about endlessness is only endlessness itself. In a finite class one can make a selection; what about infinitely many subclasses?

If mathematics were the natural science governing infinite extensions of which we can never have exhaustive knowledge, then a question in principle undecidable could be conceivable.

Wittgenstein claims that proofs work inside the system of language. To surmise what we may be able to prove, we experiment with samples. By trial and error we decide what may be true. A proof must be transparently announcing what it shows.

Using mathematics

Did not the contradictions make Frege's logic useless for giving a foundation of arithmetic? Yes it did. But then, who said it had to be useful for this purpose?

Mathematics ought to deal in facts to be useful. Mathematics teaches not answers to questions but a whole language game of questions and answers. Do these facts have to be mathematical facts? They create the forms of facts. Mathematics teaches us to operate with concepts in a new way.

Wittgenstein regards mathematics as a language game which teaches us the methods of dealing with formal systems consisting of signs. These methods go beyond mere mathematical systems, and hence the ensuing formal symbols are useful in applications. They pay by being used.

7

PHYSICS INTERFACE

7.1. Tractatus and the classical world

Classical physics is based on the assumption that there is a world constituted by certain objects, whose existence and attributes do not depend on humans and their activities. This reads to my mind as the consequence of research programs based on Wittgenstein's *Tractatus*. To bring out this point clearly, I have taken the liberty to paraphrase this work so that it can be directly applied to such an interpretation. The result is offered below.

I have chosen to retain the term "world" here, even if its absolute meaning remains somewhat ambiguous. The numbers in parenthesis after each of the adapted writings from the *Tractatus* refer to the corresponding numbers in the work. I have inserted comments of my own.

The physical world is all relations that prevail. (1)

The world is the totality of relations, not of objects. (1.1)

The world is determined by its relations, and by their being *all* prevailing relations. (1.11)

For the totality of relations determines what does prevail, and also whatever does not prevail.(1.12)

The relations in physical state space constitute the world. (1.13)

The physical world is describable by its relations. (1.2)

Something can prevail or not prevail while everything else remains the same. (1.21)

What can be established—a relation—is that a configuration prevails. (2)

A configuration consists of relationships between physical objects. (2.01)

It is essential to objects that they should be possible constituents of configurations. (2.011)

Note: Objects are entities constructed from the empirically obtained facts. Here we define the world as a configuration of prevailing relations, i.e. the typical measurement situation where a property is gauged against a standard. The configuration of all such relationships constitutes the world. This can be defined by stating which relations prevail and which do not. The world is to be defined in terms of logically independent relations. This opens the introduction of classical "possible world" descriptions. The ontology is left open by stating that things are such objects that can be parts of the relations. I now continue by citing only parts of sentences.

If an object can be part of a configuration, this possibility must be in it from the beginning. (2.012)

If I know an object, I also know all its possible occurrences in configurations. (2.0123)

If all objects are given, then at the same time all *possible* configurations are also given. (2.0124)

Objects contain the possibility of all configurations. (2.014)

Objects are simple. (2.02)

Objects make up the reality of the world. That is why they cannot be composite. (2.021)

Note: Objects are the entities found to prevail in the empirically obtained material. When they are known, we know which configurations they can describe.

Reality *can* only constitute an empirical structure, and not any essential properties. For it is only by means of formal statements that essences are described - only by the configurations of objects that these are constructed. (2.0231)

There must be objects, if the world is to be describable by a consistent formalism. (2.026)

Objects persist, endure; the configurations change, they are accidental. (2.0271)

Note: Reality must have some sort of existence to allow its description by theories. The objects, however, cannot be determined by their essence but by their possible relations. The ontology of reality determines what can be the case, but not why the objects are what they are.

In a configuration the objects fit into one another like the links of a chain. (2.03)

Formalism determines the possibility of configurations. (2.033)

The totality of possible configurations also determines what configurations do not prevail. (2.05)

Reality is the prevailing or not of configurations. (2.06)

Configurations are independent of each other. (2.061)

The totality of reality is the world. (2.063)

We make theories of facts to our selves. (2.1)

Note: "Facts" is used to denote the observable impressions which we perceive directly, possibly through some scientific methods. Facts cannot be deceptive but exist uniquely.

A theory represents a situation in physical state space, which configurations prevail and which do not. (2.11)

A theory is a map of reality. (2.12)

In a theory, the symbols of the theory represent objects. (2.131)

What constitutes a theory is that its elements are related to one another in a definite way. (2.14)

Note: Here the intentionality of our theoretical activities is taken care of.

The feature that the symbols of the theory are related to one another in a definite way represents that the objects are related to one another in the same way. (2.15)

That is how a theory is attached to reality; it reaches right out to it. (2.1511)

It is laid against reality like a measure. (2.1512)

If a map is to be a theory, it must have something in common with that which it maps. (2.16)

A theory can map any reality whose formal structure it has. (2.171)

A theory cannot, however, map its formal structure, it displays it. (2.172)

What any theory, of whatever type, must have in common with reality, in order to be able to map it - correctly or incorrectly,- in any way at all, is the structural form, i.e. the form of reality. (2.18)

A structural theory can map reality. (2.19)

A theory has structural form in common with what it maps. (2.2)

A map represents a possible situation in physical state space. (2.202)

A theory agrees with reality or fails to agree; it is correct or incorrect, true or false. (2.21)

What the theory refers to, it represents independently of its truth or falsity, by means of its formal structure. (2.22)

What a theory represents is its sense (=*Sinn*). (2.221)

The agreement or disagreement of its sense with reality constitutes its truth or falsity. (2.222)

From the theory only, one cannot decide whether it is true or not. (2.224)

There are no theories that are true *a priori*. (2.225)

A name identifies the object. The object is its meaning. (3.203)

A statement can determine only one physical situation; nevertheless it can only have sense if the whole configuration space is given. (3.42)

A statement in physics is a model of reality as we comprehend it. (4.01)

A statement *shows* how the facts stand *if* it is true. and it *claims that* this is how the facts stand. (4.022)

Statements can represent all of reality, but they cannot represent what they must have in common with reality to be able to represent it - structural form. (4.12)

Statements *show* the structural form of reality. They display it. (4.121)

What *can* be shown *cannot* be stated. (4.1212)

The simplest statement, the basic statement, claims that a configuration prevails. (4.21)

A criterion for a statement to be basic is that no basic statement can be contradicting it. (4.211)

The test conditions of a statement determine the leeway it leaves open to empiricism. (4.463)

We have said that some things are arbitrary in the symbols we use and that some things are not.

In logic it is only the latter that express. (6.124)

These modified statements constitute a program for scientific research. It applies to a classical world, which has presently been replaced by the much less concrete modern physics. Also Wittgenstein had to modify his thinking about philosophy. These modification will below be compared to those in physics. Before that, however, we need to look into some special aspects of the situation.

7.1.1. The binary description of reality

The work by Stenius [93] is an excellent analysis of the world described by Wittgenstein in the *Tractatus.* Of special interest to us is his discussion of the character and extension of what Wittgenstein terms "the logical space". According to our interpretation, this is the "state space" in which the positions denote the state of the physical system under consideration.

Stenius discusses the question whether the world can be described in terms of "yes-no" statements only. This is an interesting issue, because it seems that if a statement A is true in a "world of facts" it must correspond to a true "state of affairs" in the "world of things", which is taken to be reality. But if A is true, its negation must necessarily be false, and thus the two statements are not independent, which is required of Wittgenstein's basic facts. Thus Stenius concludes that the world must be described by the yes-facts only; they can be taken to be independent. The world is defined by the logical product of all that is true, a kind of logical conjunction of facts. But then we encounter the problem of negative facts; contrary to what Wittgenstein believes there are then no negative facts. Stenius solves the dilemma by not allowing the negative facts in the "world of facts" representing our theory but allowing them in reality.

Here the question arises whether all statements have meaningful negations? If A is an empirical statement, to claim it to be true, we must have administered a test to be certain of this. Do we have a test to validate the negation always? Knowing the extension of a concept, we may not necessarily know the extension of its denial.

In his article "On Logical Form" from 1929, Wittgenstein argues that there can be no yes-no description. Even if he later renounces this article, this opinion seems to herald his later approach to logical descriptions.

All this has acquired novel interest because of the emergence of quantum information considerations. Here it is commonly assumed that every physical system can be described in terms of binary variables. The quantum properties then correspond to projectors, which give yes-no type binary answers. The use of such states is well known and widely used in quantum information contexts.

7.2. Quantum physics turns to language

The formalism of quantum mechanics reminds one of the Wittgensteinian discourse. This is best brought out by Aage Petersen [42], who tries to combine the Copenhagen view of physics with the philosophical tradition. Here it should be noted that even the conception of modern physics has to retain features of uncritical realism. Honner writes about Niels Bohr's transcendental attitude [29]:

> Bohr is certainly open to criticism here, for his argument seems philosophically unsophisticated. He takes it as read that ordinary language can be used to provide a sufficiently objective account of experience and activity in the real world. Further, he appears to assume that any successful communication must employ terms which are related to such ordinary language. Any relativity in such language is given sufficient success by the reality of nature, *et voilà tout.*

The results of such considerations would concur with Wittgenstein's "facts" which represent contingent truth, which cannot be meaningfully denied, they persist. Only their role in human experience may be modified.

Niels Bohr always stressed the role of the observer as the actor in the drama of physics. We are not merely contemplating the world; we are acting on it, and thereby modifying its course.

Our mental picture of the world is related to reality as a map to the terrain itself. Theories are thus maps guiding us in our journeying through the world. The observed events can be used as signposts on these routes. However, an independent reality in the ordinary physical sense can neither be ascribed to the phenomenon nor to the agencies of observation. What properties persist objectively cannot be surmised from quantum theory. It surpasses ontological considerations. The use of abstract mathematics eliminates the need to fix what there really is.

The quantum mechanical approach to knowledge is adapted to man's method of framing questions. According to Bohr, quantum mechanics sets not only a limit to our observations but also a limit to possible meaning attached to the information acquired. The concept of ambiguity has extended its domain of validity from values of physical variables to the concepts constituting the variables themselves. All theoretical models thus contain elements of arbitrariness. This allows us to extend epistemological lessons to ontological images. According to Bohr:

> As for the reality concept, its logical elements seem to be related to the element of arbitrariness in physical description, i.e. to those elements of the description that are not fixed by the scheme itself, but which must, at least at present, be specified without using an algorithm. The occurrence in the mechanical formalism of elements that are not fixed but essentially arbitrary, and the necessity of fixing these elements prior to the generation of specific predictions, seems to require conceptual references to something "outside" the algorithm. It may be part of the logical function of words like "nature" or "experimenter" to contain such a reference.

Closed theoretical frameworks do not enforce their own interpretation. In physics we need to fix the "correspondence rules" to be able to apply the theory to experiments.

The mobility of the partition between subject and object allows us to include abstractions like ourselves into the discourse. In many situations, for example in introspection, experiences are inseparable from the circumstances under which they occur, and what we say is thus highly sensitive to the placing of this partition. The use of concepts like "consciousness" hinges on this.

The common language of physics is the one used by Newton and Maxwell. What can be unambiguously stated must be formulated in this language. This guarantees the consistent description of the phenomenon; the descriptive tools for a closed system. The domain of such closure coincides with what can be represented by classical concepts. What we have learned must have closure.

Quantum physics does not allow deterministic predictions. But if there is no partition between object and observation, the issue of "determinism of Nature" does not arise. The arbitrariness of our model of phenomena prevents us from predicting the outcome of an individual experiment. Only the probability remains objective, but assigning it to the separate elements of the phenomenon lacks sense; its meaning resides in the phenomenon itself.

There is no mechanism or physical model consistent with the formulation of quantum mechanics which can account for the fact that events actually occur. Part of Nature's logical function is to express the fact that, on the basis of the quantal algorithm, the outcome is undecidable but that a decision will, nevertheless, be made.

Language is a function of reality, in which reality is the independent variable and language the dependent variable. The possible structure of being is reflected in the structure of language. Concepts are tools to represent matters of fact. Language is a picture, a map or model of reality. Language is the only tool we have to intuitively capture the possible structure of reality [42].

We are suspended in language in such a way that we cannot say what is up and what is down.

Another eminent scientist who made a lasting impression on present mathematical physics, Hermann Weyl, emphasizes the central role of language in founding our symbolic description of empirical reality. He writes [68],

In living language man has tried to make his understanding explicit and communicable. I shall once more try to illustrate our basic view by an analogous situation in the realm of language. Out of our everyday language, which has often been denounced as a horsedealer's language, there has developed on the one hand the perfectly rational symbolism of mathematics, on the other hand poetry. In the living language both are contained, the poetic and the rational element. But people's living language is neither tarnished poetry nor a blurred substitute for mathematical symbolism; on the contrary, neither the one nor the other would and could exist without the nourishing stem of the language of everyday life, with all its complexity, obscurity, crudeness and ambiguity.

Even if Weyl disavows the applicability of Bohr's complementarity to his approach, one cannot help making the connection. Weyl was, of course, fully aware of the situation in quantum theory.

7.3. Wittgenstein and physics

7.3.1. Science and method, method of science

Wittgenstein is very sparingin referring to physics. However, interpreting his thinking about cognition as a universal approach to epistemology, we can take his assertions to concern physics. Here I collect relevant statements from several of his published books, which are listed with the referenses. Relations to modern physics are pointed out or taken to be seen directly.

Wittgenstein admits that the dialectical method is very sound and a way in which we do work. His discovery is that one is able to stop philosophy when one wants to. Problems are solved, not a particular problem. Analyzing philosophy by individual examples means that the presentation can be broken off.

How can a proposition be understood? Understanding is a condition for being able to apply the sentence. Do we know all the grammar of a word when using it? The place of a word in grammar is its meaning.

The grammar of a language does not come into existence until the language has been spoken by human beings for a long time.

"Understanding" is more or less interrelated processes against a background, or in a context of facts of a particular kind, namely the actual use of language learned.

What interests us in the sign, the meaning that matters for us, is what is embodied in the grammar of the sign.

We must not confuse signs with symptoms.

In [78] there is a discussion between Friedrich Waismann and Wittgenstein. This concerns the role of logic in physics more explicitly than in other writings.

Waismann: The meaning of a mathematical concept is its mode of use. The sense of a mathematical proposition is its method of verification. Propositions of grammar are like the propositions in physics and not of the same sort as the "primary propositions" which treat what is immediate.

Propositions are verified; what higher court is there to appeal to in order to find out if it is really true?

Wittgenstein: [80]: The logical structure should allow correct analysis of phenomena, i.e. *a posteriori* not *a priori*. It would be surprising if actual phenomena had nothing more to tell us about their structure. Sense data and physical objects are entirely different. Language is being used—whether it uses words for physical objects or sense impressions. How do you know what object you gave a name to?

We can postulate only rules according to which we propose to speak. We cannot postulate states of affairs. In physics we ask for the meaning of a statement in terms of its verification. The meaning of a symbol is its place in the calculus.

Wittgenstein claims that the difficulty with physics is that the statements of physics are mixed up with rules of grammar. This implies that we mix up semantics with syntax, which is a distinction not made clear in Wittgenstein's writings.

Wittgenstein agrees with contemporary physicists when they say that the signs in their equations no longer have any meaning and that physics cannot attain to such meanings; one must stay put with the signs. But scientists do not see that the signs have meaning in as much as—and only in as much as—observable phenomena do or do not correspond to them, in however circuitous a manner.

The probability of a hypothesis is a measure of how much evidence is needed to make it profitable to throw it out. A probability is neither confirmed nor verified by an outcome.

Propositions about the future are not really propositions. They govern a grammar of propositions which are asymmetric with respect to "know".

Wittgenstein ponders, what a curious attitude scientists have: "We still do not know that, but it is knowable and as a matter of time we will know it." It may be a delusion that all truth will ultimately be known.

Wittgenstein tries to clarify the relation of formal systems to scientific analysis. We choose how to speak about features of experience concerning reality. Such observations are termed facts. They are manifest and not affected by our choice of symbolic description. He questions the possibility that progress eventually will lead to a final theory of everything.

7.3.2. Physical objects

Do not think that "physical object" is to distinguish one kind of object from another. The grammar of propositions about physical objects admits a variety of evidence for every such proposition. What is eternal and important is often hidden from man by an impenetrable veil. He knows there is something under it but cannot see it. In modern physics, we cannot reach the "things themselves".

The propositions joined by "and" are not independent of each other, they form one picture and can be tested for compatibility or incompatibility. The full description of an object is not possible because it does not tell us what is essential for its existence.

When are two circles of the same size distinguishable? Logical concepts are such that their essence implies their existence. In objects of modern physics, this does not hold: Two electrons plus two electrons is not the same as four electrons.

Properties are not ingredients of the things having them; i.e. the red cannot be extracted from a red entity. The sense in which an image is our image is determined by the way in which it compares with reality. This we may call the method of "projection".

The existence of the world is not empirical because we cannot claim it to be otherwise. We interpret experience as the shadow of a fact. Nothing is gained by assuming a shadow between the experience of our thought and the reality concerned. There are two kinds of worlds: a mental world and a physical world. The physical world is gaseous and ethereal. Sentences are often used on the borderline between logic and empiricism so that meaning changes back and forth counting now as nouns, now as experience. The "uncertainty" relates not to the particular case but to the rules of evidence.

An instance of the picture of a "particle" can be extremely misleading—i.e. when applied in such a way that the picture is of no use. We may question the solidity of a floor consisting of atoms. The picture of a thinly filled space has been wrongly applied here.

If we give names to spatial objects, our use of such names depends on a criterion of identity. The old logic contains more convention and assumptions about physics than has been realized.

The subject-predicate form serves as a projection of countless different logical forms.

It is senseless to ascribe a property to a thing if it is defined to have it. The thing does not possess the property but has been selected by testing for this.

In the course of a scientific investigation we say all kinds of things, we make utterances whose role in the investigation we do not understand. We pass automatically from one thought to another according to rules we have learned.

And, when the time comes to survey what we have said, we have to clarify our thought processes philosophically.

Physics differs from phenomenology in that it is expected to establish laws. Phenomenology only establishes possibilities. Phenomenology would be the grammar of the description of the facts on which physics builds its theories. To explain is more than to describe, but every explanation contains a description. A "thought experiment" is no experiment, it is a grammatical investigation.

I cannot use language to get outside language. Nor can we get outside the evidence.

Wittgenstein tries to clarify the connections and differences between the facts, the experience, and its symbolic representation. The term "projection" is here used for what in physics is often termed "rules of correspondence". In particular, the character of the physical objects is ephemeral. Wittgenstein's ontology is far from clear.

7.3.3. Language and symbolism

That one empirical proposition is true and another is false is no part of grammar. What belongs to grammar are all the conditions necessary for comparing the proposition with reality. That is all the conditions necessary for the understanding (of the sense).

The sentence "p is possible" means that p has sense. The impossible cannot be tested. Grammar allows us to formulate true and false propositions. Thus it tells us something about the world. Thought must have the same logical form as some part of reality.

We must be able to do as much with language as the facts which it expresses. Part of a hypothesis fixed from the beginning is a "postulate". We cannot postulate the behavior of facts, but only the way we express ourselves. Wittgenstein: The relation of proposition to fact is an internal relation.

We have only one language and it is our everyday language. We need not invent new language or construct a new symbolism. Ordinary language is all right. For certain purposes an artificial symbolism can be very useful. This corresponds closely to Bohr's view of quantum physics.

A system of language can only express expectations if it can express the present state of affairs. Mathematics and logic are part of the apparatus of language, not part of the application of language. Confusion is caused by treating mathematics as a kind of natural science. Used in connection with physics, it is obvious that it differs from natural science.

We tend to look at language as a calculus, not because we do not know an exact definition but because there is none. Language as a symbolism within exact calculus follows models from science and mathematics. Only seldom does language confer such standards of exactness. Even to have expressed a false thought boldly and clearly is to have gained a great deal.

Words have those meanings which we give them; we give meaning by exploration. Many words do not have a strict meaning. But this is not a defect. Somebody knows it, but what is the good of that—if his knowing does not prove its worth in the regular course of time? Then he is no different from someone who guessed the cause. Knowing interests us only within the game.

Our ordinary language, which of all possible notations is the one which pervades all our life, holds our mind in a rigid position. Our mental cramp is loosened when we are shown a notation which fulfils our needs. These needs can be of the greatest variety.

For someone not knowing about radio receivers, a diagram thereof will be a jumble of meaningless lines. When acquainted with the apparatus and its function, he will find the drawing to be a significant picture. If you exclude the element of intuition from language, its whole function collapses. Experience decides whether a proposition is true or false but not its sense.

Whenever we interpret a symbol one way or another, the interpretation is a new symbol added to the old one. Mathematical propositions containing a certain symbol are rules for the use of that symbol, And then these symbols can be used in nonmathematical statements. It is as if our theoretical concepts provide a scaffolding for facts.

We need new concepts and we continually resort to those of the language of physical objects. The word "precision" is one of these devious expressions. In ordinary language it refers to a comparison and is then intelligible.

A hypothesis is coupled to reality—with varying degree of freedom. In the limit of no connection, reality can do anything it likes; in this limit the hypothesis is useless. Propositions about reality have a sense if they tend to agree or disagree with our experience, at least some facet of it.

Signs have a meaning if observable phenomena do or do not correspond to them. A phenomenon is not a symptom of anything, it is the reality. The phenomenon verifies itself. Niels Bohr could not have said this better:

What can be formulated as a scientific proposition can be true or false. Thus the symbolic language of science must convey information about reality. The scientist works with terms without fixed meaning and the fuzzy borders of applicability. The methods of science are intuitive and conditioned by the social nexus of the state of scientific society. To clarify the situation philosophy is called for.

7.3.4. Phenomena

Wittgenstein concludes that some notions exhaust the phenomenon and cannot display more than one attribute; examples are numbers and colors. Different numbers cannot refer to the same collection of things. Neither can different colors coexist. Are they thus "complementary"?

When we analyze phenomena, we have to decide what the "elementary propositions" are. They need not have any similarity with ordinary propositions.

Thus the concepts of modern physics do not coincide with those of classical physics.

It is impossible to extend a system; many propositions in the extended system lack meaning in the original system. Thus it is not just a "completion". Consequently an extended theory like relativity does not contain the same concept of space-time as Newtonian physics. Adding complications makes for a fresh game.

We may alter a phenomenon by altering its conditions. To do this, we need to have an idea of causal interference applied to this phenomenon.

Wittgenstein asks: Why should there not be a physical law connecting a starting and a finishing state of a system, but not covering the intermediate state? One whole experiment in which you see the beginning and the end in one figure? This reminds one of the "wholeness" characterizing quantum phenomena.

One may try to look at this as a picture of what an experiment would be if it gave this result. The picture of an experiment is not an experiment. In quantum physics "counterfactual" assumptions are not allowed in the arguments. Only what happened did happen.

How can a physical proposition be verified in several different ways. Wittgenstein: This is common in many fields. We may have different "symptoms" of the proposition. If there are symptoms, there must be some independent verification, otherwise the word "symptom" is meaningless. To apply a criterion is a tautology, to apply a symptom is a hypothesis.

Life as a weave is varied in a multiplicity of ways - concepts are not for use on a single occasion. One pattern of weave is interwoven with many others.

A physical statement is always a hypothesis. Phenomena can be different aspects of the same hypothesis. Physical hypotheses are constructed so that they connect a great number of experiences of different kinds. The value of a physical theory grows when it covers a larger realm of empirical experience. Cause is a hypothesis based on instances. Reason is not a hypothesis. We can conjecture a cause but we may know a reason. This is a grammatical use of "can".

A proposition can be verified, a hypothesis only supported.

There is no such thing as systematic phenomenology but there are indeed phenomenological problems. There is a temptation to believe in phenomenology as something between science and logic.

A pragmatic criterion for truth is usefulness in practice.

To describe an experience we need to make clear what we postulate. The elements of the description are defined by criteria, which we may choose. Wittgenstein envisages the description of an experiment from the beginning to its final end. This is exactly how Bohr describes an experimental situation. A proposition may be verified as true or false. A symptom suggests what may be the case, thus a verification may be possible. To impose some condition from the symbolic theory of physics is to introduce a hypothesis. This may be rather singular or encompassing a large area of experience, the larger the better. The nexus of all hypotheses constituting a world description may be considered a phenomenology. This related Wittgenstein's view to philosophic pragmatism.

7.3.5. Experiment

We introduce continuously new methods of measurements and so continuously change the meaning of "length". When methodology talks about measurements it does not state the optimal method for the measuring stick, even if this is part of the method. Humans alter their concepts when they learn new facts. Measuring with a stick means getting the same result with all sticks. If it did not, our whole system of measuring would collapse. There is, however, nothing mysterious about blurred intervals. "Heap of sand" is a concept without sharp boundaries. Make the smallest heap that you would still call a heap!

A system of propositions is laid against reality like a ruler. Thus we test only whole systems not individual propositions, A description is verified or falsified by a comparison with reality.

A hypothesis has a correlation with reality which is looser than that of verification. A hypothesis is not definitely verifiable. Its relation to reality is different from verifiability. The words "true" and "false" are inapplicable or have a different meaning. A hypothesis is a law for forming propositions and expectations. Describing phenomena by means of a hypothesis of a world of material objects is unavoidable. It offers a simplicity as compared with the unmanageably complicated phenomenological picture.

A physicist does not test the correctness of his deductions by an experiment but by the truth of his hypotheses. A natural law cannot be verified or falsified. It is merely probable. The sense of a physical statement is not captured by observation. Physics statements are not exhaustive.

Calculations *a priori* may agree with experience; this does not verify the calculation but the natural law constituting the basis for the calculation; and this is *a posteriori*.

A contradiction can only occur among the rules. A contradiction among the axioms of a theory makes it impossible to translate them into observations.

I know a physical proposition only when I know a method of "projection" that correlates propositions with numbers. The equations of physics can neither be true nor false. Only findings of verification can be true or false. Physics prophesies.

Wittgenstein accepts a measurement to be comparison with a standard. To be valid a measured concept must refer to a stable standard. Experimental evidence is only asserted against a background of a whole system, a relevant language game. The experiment can only be interpreted when we know the rules for relating the symbols to features of reality, the rules of correspondence.

8

PHILOSOPHICAL CONSEQUENCES

8.1. The right to freedom

There is an essential difference between the epistemology and ontology of a theoretical description. The epistemology must devise a test for the assertibility of a statement. Here full instrumentalism reigns. The totality of empirical experience may then be used to posit a formal structure, a theory, but this is necessarily underdetermined. From finite sets of facts no infinite theory can be derived. However, the ontology assumed has to be adequate to fit the facts. But it will always contain a degree of arbitrariness, i.e. wishful thinking. Thus it is important to see which predictions it makes for phenomena not encompassed by the original experience. If it fits, the theory is gaining support. As long as this works, the theory becomes more and more established. When a contradiction is encountered, the theoretical framework must be adjusted or extended. Sometimes more, sometimes less radically.

The epistemology allows us to transfer empirical manifestations of reality into knowledge. Only well-tested and provably relevant statements can be trusted. Concepts without significant consequences are of no use. This relates the epistemology to *pragmatism.* But even the best epistemology is not able to tell what there is. No rule for measurements implies existence or its negation. We are still free to assign to reality whichever entities that we think underlie our discourse. As physicists we have to assume that something exists. How else could we understand that a theory can be proved wrong? We cannot learn from experience that something does not exist. If it does not exist it cannot teach us anything.

Gödel's theorem justifies the use of abstract formalisms in natural science. Namely, if all well-formed sentences would be assigned a definite truth value from within the formalism, there would be no room for contingency. We could not formulate well-defined statements the truth value of which we may determine by experience.

The operational meaning of a theory may be tested by an assembly of operations. These can be denumerable at most. Thus they fit into a formalistic network where the testable propositions are Gödel statements. But any sequence of empirical outcomes can only condition the theory in at most denumerable ways. According to a theorem by Löwenheim and Skolem such a denumerably verified theory can derive from a multitude of nondenumerable models. But these models may presume different entities to exist and thus they are supported by different ontological commitments. As no fully complete theory of reality is expected to exist, we will never know "what there really is".

This situation allows many possible schemes for human existence. Many social and emotional systems can coexist with our scientific knowledge. Ontology may contain many things beyond the reach of the empirical totality. Examples: Laws of Nature, consciousness, ethical values, gods, etc. These may not be subject to full empirical verification.

8.2. Pragmatism at stake

8.2.1. Summary of pragmatism

In the earlier chapters, I have presented the opinions of Boltzmann, Bohr, and Wittgenstein in a manner indicating their close adherence to the pragmatic thinking of James. To stress the point, I summarize the main tenets of *pragmatism* as it is usually understood:

1. All conclusions about the world must be based on experience.
2. Only such conclusions which have practical consequences are to be retained.
3. Our experiences are not true or false, they just come and go.
4. Only what has proved its value in practice is assertible.
5. Only that is real of which we have to take account.
6. Truth is what pays.
7. We ought to believe in the accumulated teachings of our time. Its tenets are to be trusted, but can be changed in view of new experience.
8. Conceptual knowledge is inadequate to capture all aspects of both reality and formalism (mathematics!).
9. We can never get a complete understanding of reality.
10. One may question the complete causality of physics and the assumed continuity of change.

These statements are chosen from the writings of James, but they could have been extracted from Bohr or Wittgenstein. Also as seen above, Boltzmann anticipated many of these points. Thus the summary of opinions termed pragmatism captures a genuine spirit of the time. Here I leave out the application of "pragmatic truth" to moral questions and guides for conduct. Their interpretation and application vary in the case of each person, mostly due to the individual personalities involved.

The other father of American pragmatism, Charles Peirce (1839–1914), is more careful in defining the concept of "truth". He defines this as the final view eventually emerging asymptotically when the scientific and social processes proceed. As noted above, Wittgenstein explicitly questions the value of this opinion. As things stand at present, there are no strong reasons to trust this expectation to become realized.

The moral aspects of pragmatism are brought out by the philosopher John Dewey (1859–1952), who like Peirce also contributed to the development of formal logic.

The pragmatic attitude of Bohr has been stressed by Murdoch [30]. He points out that the pragmatist is close to the positivist in his opposition to metaphysics. The positivist admits the existence of what can be observed. The pragmatist considers what he can do. Murdoch expresses this as follows:

> To attribute a property to an object, for whatever reason, is cognitively meaningful only if the attribution "makes a difference" so to speak i.e. only if the attribution implies some definite and discernible practical effects; if it generates no predictions about the future, it has no practical consequences, and hence no cognitive meaning,

Murdoch also traces the idea of fragmentation of human knowledge back to Kierkegaard's multi-aspect view, i.e. reality can be comprehended only from different points of view, in terms of disparate conceptual schemes. Bohr's world consisting of mutually exclusive physical phenomena and Wittgenstein's language games may well derive from developing this view within their comprehension of human experience.

8.2.2. Pragmatism in the modern world

In spite of its age, pragmatism is far from dead in the modern philosophical debate. Accompanying the failure of the logistic movement to provide a secure foundation for a description of the world, pragmatic criteria for meaning are really alive. Only they remain, when both the edifice of formal mathematics and the ultramodern physical theories crumble into fragments.

Hilary Putnam [101] has considered the actuality of pragmatism incorporating this into his own attitude to realism. He writes:

> There is no "Totality of observational facts" fixed in advance to be described. What is there, even at the level of observational facts, will depend partly on what culture we create, and that means what language we create.
>
> Testing predictions of a theory makes sense only when a common world and a common language are already in place.
>
> Pragmatism is "antisceptical" but "truths" are fallible and need revision. Doubts need justification as well as beliefs.

Also the influential philosopher Willard van Orman Quine (1908–2000) supports a pragmatic approach [102]:

> Hence it is meaningless, I suggest, to inquire into the absolute correctness of a conceptual scheme as a mirror of reality. Our standard for appraising basic changes of conceptual schemes must be, not a realistic standard of correspondence to reality, but a pragmatic standard.

8.3. Husserl's last stand

8.3.1. Life and fate

Edmund Husserl was born in 1859 and lived to 1938. He seems to have taken on himself the burden of the great system builders of the nineteenth century, Kant and Hegel. He wanted to capture reality in a solid framework based on intense philosophical investigations. He tried to identify the elusive "things in themselves" and capture them in a network of realistic concepts. His publication career stretches from the early investigations of mathematics in 1887–91 to the great final works published only in 1952.

Husserl struggled with his chosen subjects and turned over his earlier convictions by modifying and extending his technical terminology. This makes it heavy to penetrate his thinking, and the changing perspective does not allow one to fix his ideas and opinions. In the present context, it would take us away from the main argument to attempt a summary of his achievements. Only as a contribution to the philosophy of science do I consider his work. Beyond that the task exceeds my knowledge of his contributions and their influence on Continental philosophy.

Husserl is known to have read James's *Principles of Psychology* in 1894. He was thus presumably aware of the pragmatic turn of philosophy. But I see his work as an attempt to provide a solid foundation for theories and their relation to the real world. The attempt has turned out to be futile. The development of modern physics has made his central concept of an object obsolete. The world cannot be described as a collection of well-defined things as realized by both Wittgenstein and the quantum physicists. The ephemeral quantum systems do not conform with the classical concept of an object. Wittgenstein came to realize this in terms of linguistic shortcomings. Concomitantly the same conclusion was reached by Bohr, when he contemplated the role of human communication in modern science. Consequently, Husserl is not taken seriously in the modern philosophy of science.

The philosophical effort has been to analyze the structure and practice of linguistic communication. Following this, the researchers have tried to apply the ensuing theories to the description of the world where we live. Sense data, logical constructions, and empirical phenomenology have been tried as the founding framework for the construction of a mirror image of reality. However, the evasive experience of modern physics has made all such attempts fruitless. Furthermore, the multifarious experience of a human life fails to fit the folds of theoretical investigations. All that remains is Bohr's closed phenomena and Wittgenstein's language games. Bits and pieces of reality can be grasped and depicted but the totality remains elusive.

Following Husserl, the Continental philosophers turned their backs on the realities of the scientific endeavor and took up the investigations of human linguistics as an autonomous realm of research. What humans say, write, and exclaim becomes the material which is subjected to intense deconstruction. This touches reality only in its connections to sociology and human reactions to words. Husserl claims that Heidegger has turned phenomenology into sociology. Sometimes such concentration on the text as a text leads to churning idly a formal apparatus created for that purpose. In view of a hope for a scientific philosophy, this seems to be the fate of much *existence philosophy*.

8.3.2. Husserl's main ideas

Husserl regards truth to be objective and phenomenology the tool to bring this into the open. The investigation has to be "pure" and independent of any human opinion about cognition. The science of pure meaning cannot rely on the ordinary use of words, but must go back to "things in themselves". "Phenomena" are directly appearing to our intuition. Furthermore any vague common understanding of Being may be strongly infiltrated with conceptual theories and opinions about Being. Meanings have relations to other meanings in order to form a network. Phenomenology is a descriptive science of consciousness in general. It is a general philosophy of essences.

Husserl starts from mathematics, but he develops a phenomenology, where each emergence is nothing but its impression on the subject. No genetic or theoretical frame imposes itself on the experience. It is to be comprehended in its pure being and carries no implication about the existence of facts.

The way we apply our phenomenological concepts to real objects is termed *intention*, a term borrowed from medieval scholastics. To purify the object of phenomenological discourse, one has to apply what Husserl terms "bracketing" (*epoché*), which is supposed to free the phenomenon from its standard references to the world. In this mood, the influence of the external world fades away and only the ego with its categories remains. However any Being carried by categories of cognition indicates a mode of potential empirical existence. What is experienced through our intuition is to be comprehended but only as it offers itself directly.

The world is accidental and only the awareness of the pure ego carries meaning. The transcendental world is totally conditioned on the individual awareness. All real entities are constituted by the intellect. Such an intellect presupposes the existence of consciousness which assigns it meanings.

The question is: "How does the object come into the epistemic possession of the thinking person?" We can use ordinary experience and ordinary language but not any scientific or logical principles. Phenomenology must remain pure,

Knowledge has to be based on experience not on conventions. To know something is to be able to recognize it. The person posits ideal objects as belonging to "reality". In addition to physical objects, there are objects like "4" and its ilk. They can be the subject to false or true statements,

Husserl introduces the concept of "eidic science" to eliminate the dependence on accidental contingent facts. The general has to be deduced from the individual case. The only purely eidic science presently in existence is mathematics.

Phenomenology must be distanced from all natural sciences as well as from all inherently naturalistic views. Husserl believed that the self-evidence characteristic of "$12 \times 12 = 144$" could be found also in other fields of human knowledge.

Consciousness is not part of the world at all, because it is the reason there is a world for us in the first place. It does not create the world, but the world is opened up for consciousness. There is no conceivable world without consciousness. Husserl's concept of "*epoché* of reduction" is taken to lead to objectivity constituted out of subjectivity. The world is a product of a constituting ego. This ego is beyond the world. It gives the possibility to understand the world. The transcendental subjectivity within a communalized life constitutes an objective world, which is the same for everyone. The life world is between the world of nature and the world of culture.

And expressions refer to an object; not only do they have meaning but also reference (Frege's *Sinn* and *Bedeutung*). Sentences are true because states-of-affairs make for truths. According to Husserl it is in words and language that things first come into being. Husserl's act of positing a world eventually makes him leave his commitment to realism. Everything that enters consciousness can be studied by phenomenology. This leads Husserl to a version of transcendental idealism. All meanings and essences are engraved on the transcendental ego. He has also been accused of eventually becoming a pure idealist.

Husserl's methodological solipsism is not regarded as a reliable method in science. Many philosophers deny the possibility to separate the empirical from the *a priori* in our knowledge. The teaching of Husserl seems to me to be an attempt to impose a conception of consciousness on a Kantian frame of categories of the mind. The influence of scientific discoveries beyond the everyday experience makes the move unable to compete with other more modern philosophies of science. His grand enterprise to erect a cathedral of theory over the individual experience of the world could not work.

8.3.3. The crises in European science

Due to the dynamic character of Husserl's writings, any brief summary, like the one above, is necessarily misleading. A fair presentation should follow the treatment in its evolution with time. Here this has fallen outside my aim with the introduction of his thinking. As a lingering ideal from the great systems of the nineteenth century, he represents the failure of the past approaches, To state the point more succinctly, I choose to comment more on one of his latest works, *The Crises* [103] published originally in German in 1936. The editors have based the text on his writings but do not necessarily give a proper picture of his actual ideas. However, its topics are most directly related to the present investigation.

Husserl's concept of the "Crisis" does not signify any failure of the scientific endeavor. On the contrary, it derives to a large extent from its spectacular success. This, however, derives from a demarcation of science from the natural world of experience, which is the only world where a human can live.

The problem can be traced back to Galilei, who introduced the idea that nature can be understood in terms of an idealized mathematical model. This view has been perpetuated by the sciences up to the present time. All scientific results depend on replacing the naturally experienced concept by an ideal model which can be treated by manipulating the signs constituting the abstract structure. The scientific treatment is not exact but only carried out using exact terms. The result emerging may be exactly derived within the model but it can only be applied to human reality by intuitive interpretations.

The conclusion is that the original motivation for the idealization is forgotten. The historical and social roots of the science are lost, and the mathematical model is identified with reality. What cannot be observed and captured in numbers does not exist. Positivism becomes the prevalent creed, and it is forgotten that positivism itself is a metaphysical presumption. This attitude is sometimes termed "scienticism" and considered to resemble a religious conviction. It does, indeed, miss the need to integrate scientific experience with ordinary human life. What is not scientifically meaningful is illusionary and irrelevant. According to Husserl:

> In our vital need—so we are told—this science has nothing to say to us. It excludes in principle precisely the questions which man, given over in our unhappy times to the most portentous upheavals, finds the most burning: questions of the meaning or meaninglessness of the whole of this human existence. Do not these questions, universal and necessary for all men, demand universal reflections and answers based on rational insight? In the final analysis they concern man as a free, self-determining being in his behaviour toward the human and extrahuman surrounding world and free in regard to his capacities for rationally shaping himself and his surrounding world.

The solution is to recover the roots of the sciences and by the methods of Husserl's phenomenology situate the empire of knowledge in its proper place in human life. The concept of "truth" may carry different meaning in the contexts of realities such as emotional, social or moral issues. Only by situating oneself within the original historical and social framework, can the individual grasp the significance of knowledge in human life. Experience of such transcendental truths is acquired only by philosophical contemplation carried out by the individual. The result is, however, a genuinely human world which can be shared by everyone.

The full conception of the phenomenological world resides in what Husserl terms the "life world" (*Lebenswelt*). This comprises the totality of the human sensational experience melted together with all aspects of culture and science. In this world live both the emotional and the rational aspects of humans. In its multifarious conglomerate, it may well carry part of the meaning of Wittgenstein's concept of a "life form". The difference is that Husserl hopes for a unified view of the total reality.

All the occasional (even "philosophical") reflections which go from technical work back to true meaning always stop at idealized nature; they do not carry out the reflection radically, going back to the ultimate purposes which the new science, together with the geometry, which is inseparable from it, growing out of prescientific life and its surrounding world, was from the beginning supposed to serve a purpose which necessarily lay in this prescientific life and was related to its life world.

Husserl grasps the modern trend to found the philosophical world in the singular experiences of the individual, just like Wittgenstein and Bohr in their respective realms, but he was seriously in conflict with them concerning his motivation. Human consciousness and things in themselves have disappeared just through the scientific progress decried by Husserl. His philosophical method was based on essences and introspection which are not regarded as valid tools any longer. Rightly or wrongly, modern philosophy does not allow one to conclude anything about the entities behind the phenomena represented in language.

The futility of his method was eventually acknowledged even by Husserl in 1935: "Philosophy as a method in rigorous science is a dream which is over." Thus ended an ambitious and global attempt to bring the world back into an ordered shape. We have to live with the fragments that remain.

Let us turn to the final verdict on the existential movement of philosophy as seen by Hermann Weyl. He was a student of Husserl but turned away from his teachings because of his experience of modern science and its description in terms of innovatively created symbolic frames. This he saw most blatantly exhibited by Husserl's main successor, Martin Heidegger. Weyl states [68]:

Phenomenology had lunged out in all directions beyond the narrow boundaries of sensual data, but still its pure ego was like a spirit soaring above the waters, untainted by worldliness, and it was hard to conceive how it could ever surrender its immanence and become flesh, man among men.

9

METAPHYSICS AND REALITY

9.1. What can we know?

9.1.1. Embedding knowledge into reality

Each human being is a complete world. He does not experience a world, construct a world, or live a world. Nay, being aware of sense impressions, conscious meditation, introspective impulses, and the stream of thought, he is a unified entity which is not an assembly of these but the active subject determined by all of them. It is also experienced as senseless not to accept one's experience of social conditions by others as equal with the experienced self.

The human complex is the object of all that is knowable and the object for investigation by all activities worthy of the name science. By constituting simultaneously the agent recognizing regularities and correlations between various aspect of his being, the individual modifies his vision of reality, thus building up what he regards as knowledge. The characteristic expression of cognitive efforts is to present the results of thinking in the framework of language. Its use is a multifarious activity, which ascribes life's various situations to others more or less unambiguously. As the variety of situations encountered differs both in its physiological aspects and psychological significance, language can only capture a limited part of reality at any one time. The linguistic reality we live in is fragmented and incomplete, but it is capable of changing and adapting to extended experience.

By constructing technical devices and designing measurements man has enriched his potentialities to make formal models, games, and symbolic systems. These extend the tools of language and form the foundations for all our knowledge.

Knowledge is thus the formation of partial maps of experiences conditioned by our mental developments and abilities. However, when relating to the reactions of the individual person or the more exact results of measurements only correlations between observed data can be established. Linguistic analysis can only answer the question "How" but not the question "Why". The ultimate causes remain out of reach.

Living is the emergence of the human totality which seems to proceed from the biological necessity inherited through the species. No formal analysis is guiding this, but ordinarily linguistic ability develops along with the conscious personality. As the child learns to act, he learns to comment on his actions and the surrounding ambience. The formal description comes from learning abstract systems and mathematical models. This development is regularly observed but still remains the ultimate human mystery.

Thus in spite of all accumulated knowledge, our understanding is strongly limited. Why do we experience consciousness, and why can we analyze and affect the processes of microscopic physics? We do not even know the limits of our mental tools.

Even if we can describe the outcomes of any isolated physical experiment very accurately, we cannot derive the macroscopic world where our daily life takes its course. Even if we, one day, could measure all physical processes in our brains and correlate them to mental experiences, we might not be able to derive the concepts of common consciousness. Have we, possibly, encountered some fundamental demarcation between different conceptual nexuses?

The problem is that even our best scientific achievements only give correlations not causes. The activity is purely *epistemological* relating the posited entities and phenomena to each other. The actual origin and character of these is not reached by our present scientific methods.

9.1.2. On what there is and is not

The ultimate elements of our symbolic representations have not been disclosed by their success. What there really is falls outside the explanations of science. Theories contain no necessary *ontology* but stand in need of a sufficient one.

In formal mathematical frameworks, the elements are defined by their role as signs operated on by rules. Else they are nothing. Thus different models may be shown to be isomorphic in fitting all the axiomatic features of the structure.

In physics, matter consists of atoms, atoms consist of nuclei and electrons; these consist of quarks, gluons, and leptons; these consist of "???". No theory, so far, predicts the numbers and properties of these building blocks. Will we ever be able to do so?

It is a possibility that our way of approaching science is incapable of settling ontological issues. In order to proceed with science, the researcher has to assume that something exists; the activity of research must assume that the concepts developed can be asserted to apply to something, or alternatively fail to do so. The best proof of an external world is that even the best formulated theory may turn out to be incorrect. Any statement about nature may turn out to be false. Thus the best argumentation for a correspondence theory of truth is that such a correspondence may be shown to be wrong. It is not what there is that determines our image of reality but more appropriately what there is not. This is a generalization of Popper's *falsification theory*.

If the positing of an element x is defined by the truth of some proposition $p(x)$ being true, it is of central importance to know what state of affairs is excluded by the truth of $\neg p(x)$. It is essential: not only do we need to know what is the case, but even more what is shown not to be the case. This is the scientific version of the logical problem of defining negation.

For each situation that is the case, there are in reality more possibilities for it not to be the case. Thus only in Wittgestein's logic are the spaces of a concept and its negation of the same size. For anything imagined, there are more things that we cannot imagine. Reality is too multifarious to be captured by any closed system of formalism.

9.2. Beyond the veil of endeavors

9.2.1. Leeway for beliefs

I have claimed that science is epistemology but fails to provide a well- defined ontology. This allows for the challenging opinion that even if all scientific questions have been decided, there is still an unfathomable realm of concepts and relations which are fully consistent with all known. We are free to select and choose what we believe constitutes ultimate reality.

In the logical and mathematical world this manifests itself in Gödel's theorem. There are more statements that can be declared true than what can be proved in any fixed system. If we identify a statement as *undecidable*, we only need to look for an extended model of the system to make it true or false. Here we are free to choose; no single choice is necessary.

In physics our theories are not strictly logically formal, but based on heuristic and intuitive use of formalism. Still the sum total of what we think we know is a limited amount of information. It can never be complete and closed to re-interpretations of the basic concepts. And this is good: if it were not the case, science would know no progress.

So what do we have to believe about the fabric of reality? How do we choose what to found our lives on? We may want to live in a rationally constructed world, but this has usually failed to emerge. Emotional attitudes break through and may, in extreme cases, break down the illusions of one's life. Alternatively, various ideologies supply frames where one may find a model for conducting one's life. A most developed system of this type is offered by the various religions whatever their origin may be.

In the discussion about certainty carried out by Moore, Wittgenstein, and von Wright we found that all cognitive activity must be based on the received view supplied by the collected wisdom of the time and place we are born into. Only by recognizing what we presume known, can we proceed to question individual results. Something can be shown wrong only with respect to all the features of consensus teaching. Successful testing of an individual theoretical proposition also serves as a test of the whole theoretical game. Eventually enough internal conflict will lead to a total revision of the whole system: the Kuhnian revolutions are, however, rare.

A given theoretical fabric is always given by the life world where we grow up. From this our education has to proceed. What teachers tell us is to be accepted by their authority. Questioning comes only later. The concept of "knowledge as true belief" is an unattainable ideal; a better conception is knowledge as "received teaching". This better describes the limited and dynamic character of the information available to humans.

As von Wright states: "In a genuine knowledge situation there must be grounds for knowing." To know nothing is not to exist as a human being.

William James in *The Will to Believe* argues for our right to make decisions beyond the realm of life where sufficient evidence exists or is likely to appear soon. On many profound issues concerning our lives, this situation arises; we have to make decisions but we lack the factual information needed to guide us. Still, *not to decide is a decision in itself*; it may be that we miss opportunities for happiness or success.

James is closely concerned with moral questions, but I see no reason to limit his argument to these only. As we noted, all scientific progress with its successful applications fails to inform us about the ultimate essence of reality and the guidelines for our conduct as humans. Thus we must decide, but a fully rational decision is not at our disposal.

James defines a *hypothesis* as anything which offers itself to our beliefs. A live hypothesis is one representing itself as a real possibility. A hypothesis may be dead because it seems anachronistic or irrelevant.

> When we look at certain facts it seems that our passional and volitional natures lie at the root of all our convictions. When we look at others, it seems that they could do nothing when the intellect had once said its say.
>
> ... but for us, not insight, but prestige of the opinions, is what makes the spark shoot from them and light up our seeping magazines of faith.
>
> Our belief in truth itself, for instance, that there is a truth, and that our minds and it are made for each other - what is it but a passionate affirmation of desire, in which our social system backs us up.

James thus claims that many of the decisions supporting our lives have been chosen by faith not arguments:

> Our passional nature not only lawfully may, but must, decide an option between propositions, whenever it is a genuine option that cannot by its nature be decided on intellectual grounds; for to say, under such circumstances, "Do not decide, but leave the question open." is itself a passional decision just like deciding yes or no, and it is attended with the same risk of losing the truth....
>
> I have long defended to my own students the lawfulness of voluntarily adopted faith; but as soon as they have got well imbued by the logical spirit, they have as a rule refused to admit my contention to be lawful philosophically, even though in point of fact they were personally all the time chock-full of some faith or other themselves.

According to James many of the opinions and convictions we have, do not have their origin in rational evidence. From a spectrum of beliefs we have to take a "leap of faith" which fixes essential parts of our life. This does not conflict with the rational endeavor of science. As noted above, science leaves open many questions central to life. As stated by Wittgenstein:

We feel that even when all possible *scientific* questions have been answered, the problems of life remain completely untouched.

He continues: "Of course there are then no questions left, and that is the answer." However, the problems of life have refused to go away, they linger on both in natural science and moral discourse.

Consequently nothing we have learned or will learn can make all questions silent.

9.2.2. Role of religion

One of the most developed networks for beliefs is offered by religion. Its role has always been central in most human societies and in the lives of individuals. This fact neither supports the existence of God nor its negation. It just happens to be the most influential feature of human attitudes, even when only sustained by beliefs chosen or imposed. Its relation to scientific progress is autonomous; there is no positive argument in its favour, but this does not justify the often expressed opinion that there can be no God. Even sophisticated scientists have not denied the reality of religious experiences. It emerges also in the arguments by the characters treated in this work.

James on belief

James's *Will to Believe* is actually aimed at religious faith:

Now, to most of us religion comes in a still further way which makes a veto on our active faith even more illogical. The more perfect and more eternal aspect of the universe is represented in our religions as having personal form....

When I look at the religious question as it really puts itself to concrete men, and when I think of all the possibilities which both practically and theoretically it involves, then this command that we shall put a stopper on our heart, instinct and courage, and wait—acting of course meanwhile more or less as if religion were *not* true—till doomsday, or till such time as our intellect and senses working together may have raked in evidence enough—this command, I say, seems to me the queerest idol ever manufactured in the philosophic cave.

James early considered the religious impulses and their manifestations in human behavior. A report on his observations, *The Varieties of Religious Experience* [57] appeared soon after *The Will to Believe.* The subtitle of the work, "A Study of Human Nature", suggests the character of the approach. The book treats religion both as a social and a philosophical phenomenon. Among other things, he discusses what can be known about God and what we can only observe in humans. He presents some of his conclusions:

If, namely, we apply the principle of pragmatism to God's metaphysical attributes, strictly so called, as distinguished from his moral attributes, I think that, even were we forced by a coercive logic to believe them, still we should have to confess them to be destitute of all intelligible significance....

So much for the metaphysical attributes of God! From the point of view of practical religion, the metaphysical monster they offer to our worship is an absolutely worthless invention of the scholarly mind....

But what shall we now say of the attributes called moral? Pragmatically they stand on an entirely different footing....

It needs but a glance at them to show how great is their significance....

An intellect perplexed and baffled, yet a trustful sense of presence—such is the situation of the man who is sincere with himself and with the facts, but who remains religious still. We must therefore, I think, bid a definitive good-bye to dogmatic theology. In all sincerity our faith must do without that warrant.

Thus James dissects both the logical and ethical arguments for God's existence and finds them ineffective. The belief must be founded in faith, and faith is taken to be a philosophical concept.

It would be unfair to philosophy, however, to leave her under this negative sentence. Let me close, then, by briefly enumerating what she can do for religion. If she will abandon metaphysics and deduction for criticism and induction, and frankly transform herself from theology to science of religion, she can make herself enormously useful.

The spontaneous intellect of man always defines the divine which it feels in ways that harmonize with its temporary intellectual prepossessions. Philosophy can by comparison eliminate the local and the accidental from these definitions. Both from dogma and from worship she can remove historic incrustations. By confronting the spontaneous religious constructions with the results of natural science, philosophy can also eliminate doctrines that are now known to be scientifically absurd or incongruous.

In summary, James defends our right to believe in those parts of what we learn by education, which we find to satisfy our craving for supernatural existence. However, no rational argument or empirical progress can dictate what we have to believe. Philosophy can only survey the faiths and guarantee that they are situated in the realm where our epistemological machinery lost its power to direct us. What we cannot know or disprove, we may still will to believe.

Bohr on religion

In the talk "Physical Science and the Study of Religions" in [37], Niels Bohr discusses the relations of science to beliefs. His point of departure is physics and its development. His approach sounds very pragmatic, which may well derive from his contacts with James's writings.

In any attempt of clarifying man's position in existence, it is a question of the proper balance between our want for an all-embracing way of looking at life in its multifarious aspects and our power of expressing ourselves in a logically consistent manner

Typically Bohr analyzes the situation in terms of complementarity. In this case particularized to the freedom to choose the separation between object and observer at will:

In fact, the varying separation line between subject and object, characteristic conscious experiences, is the clue to the consistent logical use of such contrasting notions of will, conscience and aspirations, each referring to equally important aspects of the human personality....

This attitude may be characterized by its striving for harmonious comprehension of ever more aspects of our situation in the never ending struggle for proper relation between content and frame, recognizing that no experience is definable without a logical frame and that any disharmony apparent in such relationship can be removed only by appropriate widening of the conceptual framework.

In its context of religious discourse, this may be interpreted to say that Bohr regarded moral and divine issues as an unavoidable part of human life. It can be confined to the individual consciousness by a freely chosen demarcation against the physical phenomena whose description is dictated by the formal framework of our theories.

Bohr's attitudes to religion are thoroughly documented by Honner [29]. Most of the evidence is found in Bohr's letters and reported conversations. Hence I will illuminate the issue by directly quoting Honner.

Bohr recognized similar problems of physicists and theologians in the application of language to the extraordinary events in the sub-atomic and the supernatural....

In physics we do not talk about God but about what we can know. If we were to speak about God, we must do it in an entirely different manner.

The religious questions are outside the epistemology handled by the language we know. This, however, does not entail the nonexistence of ineffable features of reality; only we cannot reach them with our empirical way of life.

On the other hand, the same epistemology, based as it is on the limitations to known experience and language manifests an openness to unlimited and ultimate harmony....

One would never substantiate the claim that Bohr ever enjoyed a mystical experience....

There is considerable support for Bohr's acceptance of an ever wider impression of an eternal and infinite "harmony" and a coherence that in itself can never be fully described but always more deeply studied. Thus belief in a One and the Many is not just the secret of success of science, as Bohr viewed it, but also a clue to a more rational metaphysics.

One should not, however, put too much stress on Bohr's "holism". His "wholes" are the completed descriptions of physical phenomena, and he cannot be accused of considering the totality of being as a phenomenon. But there is no denying his quest for a universal harmony in our experience of reality. We may not know what really exists, but our accumulated knowledge about Nature should be complete and consistent.

When all possible scientific communication has taken place, the rest is not silence but personality and belief. Like James, Bohr does not impose any religious ideas on us, but recognizes the fact that science leaves room for them, if we choose to believe.

Early Wittgenstein

In his later writings, Wittgenstein scarcely ever refers to God as a divine entity. He is used as a *gedanken* reference in logic, not as an existing and acting concept. Only in the *Tractatus* 6.432 does he write: "God does not reveal himself in the world." This suggests the question: If his existence were something that "shows itself", where would he reveal himself?

There are, indeed, things that cannot be put into words. They *make themselves manifest.* They are what is mystical....

Feeling the world as a limited whole—it is this that is mystical.

In this Wittgenstein almost verbatim concurs with the expressions chosen by Bohr.

In the early notes before this [104], written during the First World War, the references to religion are far more common. At that time Wittgenstein, understandably enough, was more immersed in the fundamental issues of life and death. In fact, he never abandoned his devotion to moral and existential questions. His struggle with language and logic questions induced fewer direct religious references in his writings. He states in the early notebooks:

To believe in a God means to understand the question about the meaning of life.

To believe in a God is to see that the facts of the world are not the end of the matter.

To believe in God is to see that life has a meaning.

Thus Wittgenstein concludes that, in addition to the "facts of the world", human life depends on beliefs.

As for what my will is, I don't know yet.

That is why we have the feeling of being dependent on an alien will.

However this may be, at any rate we *are* in a certain sense dependent, and what we are dependent on, we can call God.

In this sense God would simply be fate, or, what is the same thing, the World - which is independent of our will.

But even this underdetermined concept of God can influence our way of living:

I am then, so to speak, in agreement with that alien will on which I am dependent. That is to say "I am doing the will of God."

Certainly it is correct to say: consciousness is the will of God.

Conclusion

We have seen that the philosopher James regards religious experience as an essential part of human life. The physicist Bohr has to take them into account when situating physical phenomena into the world of man, and finally Wittgenstein uses the concept of God to found the moral obligations of humans. One may say that their attitudes are in conformance with our earlier insight. Theories about reality are always underdetermined by factual experience. Thus even the most accomplished theory can only carry its operations so far. Beyond the formalized experience there are extended worlds which can be adjusted by the individual according to concepts and opinions formed freely by the conscious mind. It is not only possible but also necessary and desirable that we have enough personal faith to carry out our human lives.

Religion comes in as one possibility. No philosophy or science can imply the validity of religion. But it still exists in the human mind and society.

Religious convictions come in many different shapes. They may look like pieces of philosophy or politics, they may be historical or juvenile, but they control our lives. Not to speak of the emergence of very recent caricatures of religion, which have emerged from pop music or drug culture. Bizarre as these may seem, they derive from the same needs and survive in the underdetermined realm of mind. If it is not experienced that they infringe on the realm of what is stateable and verifiable, they derive from the limited knowledge of their practitioners, which can only be amended by better education.

10

CONCLUDING EPILOGUE

The present work follows the path of the philosophical and metaphysical discussion starting in the nineteenth century and carrying us into the just-passed twentieth century. Many ideas have waxed and waned in this period; the history has developed from the seemingly stable Europe emerging from the Napoleonic Wars. For a while political stability seemed to be guaranteed; the old order was to dominate in a Europe destined to rule the underdeveloped people outside of it. As we know now, all this was an illusion; two great wars shattered all stability and the creed of Marxism spread only to suffer the most spectacular failure experienced by any ideology. First European stability broke down, and then the fragments had to be combined into new and conflicting structures. These have gone through many stages of more or less chaotic transformations, and after experiencing the international disorder presently prevailing one finds it hard to guess where we are heading next.

The starting point for our analysis is the progress of science, especially physics, taking us from the Newtonian determinism to the present era with indeterminacy and counterintuitive features following from quantum mechanics. The development leads us from the fully causal Maxwellian theory through the statistical chaos of thermal physics. The uncertainty emerging in quantum theory is fundamentally different from the kinetic disorder of atomic ensembles.

My argument is based on the fact that the social and scientific security offered in the nineteenth century evaporated without discernible reason. In the human mind, this emerged in the anarchistic philosophy of Nietzsche and the realization of the existence of our subconscious impulses by Freudian investigations. We know now that we are neither the simple animals of Darwinian evolution nor the rational masters of our culture. These insights have offered the background for multifarious human conflicts treated in the literature of the past century.

The development in science has mirrored this general trend. My object of investigation is the breakdown of the logical structures envisaged by the positivistically inclined builders of well-ordered worlds. Here I choose Ludwig Wittgenstein as the chief representative of this phenomenon. His published work *Tractatus* formally seems to close the world in a logical way. Everything that can be said is to be said and the rest is silence.

As a logical construction, his work seems to capture the world in its totality, but even the very description brings up the need to go beyond the logically secure. He necessarily has to talk about what is deemed to be passed over in silence.

The possibility of a closed consistent logical framework was supposed to be demonstrated by the establishing of mathematical systems by Russell, Whitehead, and Frege. The path, however, turned out to be littered with stumbling blocks. Gödel put the final hurdle in the way of progress. The hoped-for certainty evaporated when investigated in detail.

The loss of naive understanding was even more patent in natural science. First, Einstein destroyed the permanence of space and time; then quantum physics removed the remaining vestige of the hope to capture reality in a nexus of logically introduced symbols. The extremely successful quantum theory does not provide a description of our experience as a closed totality. Modern science does not provide a world where humans can live.

The classical universe of Newton and Maxwell has been replaced by Bohr's ephemeral description of human knowledge as an epistemological construction whose relation to reality is diffuse at best. The world of quantum physics must have some existence, but its character seems to be given entirely through its reactions to human decisions.

In this book, I consider Wittgenstein's conversion from strict logical analysis to an investigator of human communication, with all the potentialities and limitations of ordinary language. This process I put up against the transformation of nineteenth-century mechanistic science into the present pragmatically oriented world image of modern physics. The prophet of this move has been Niels Bohr, who has advocated the incomplete ambience of the quantum world. His epistemological view has proved its utility, but it has regularly been questioned in many ways. However, no competitor has been able to replace it.

In this book, I have tried to map the path of the nineteenth-century certainty into the shattered world image of the dawn of the twenty-first century. This is manifest in the confusing progress of science and the disappointing attempts to order the worlds of logic and mathematics. My contention is that the conversions of Wittgenstein and Bohr are analogous reformations of the methodology of research. Both end up with the conviction that all real knowledge must be based on ordinary language working as communication device between individuals. Without language, outcomes of experiments cannot be declared publicly and no *private language* can be of value in human intercourse. I find that, in spite of no direct communication between them, Bohr and Wittgenstein make very similar declarations of their vision of the epistemic activity by humans.

Ideas presented here are closely similar to some opinions found in works by earlier writers. In philosophy, the *pragmatism* of William James is remarkably like the modern philosophy of science. But then both Wittgenstein and Bohr are influenced by him, at least to some degree. Wittgenstein was to have studied with Ludwig Boltzmann; this opportunity was lost because of Boltzmann's death. His essay writings, however, seem to foreshadow some central themes in Wittgenstein's philosophy. Remarkable support for my opinions are to be found in the essays written by the scientist and mathematician Hermann Weyl.

Along the way, we consider also a multitude of different intellectual systems, mostly finding them relevant but mainly only marginally. Even mathematics gives no final certainty, phenomenological approaches fail to deliver, and realism turns out to be doubtful. What is left is a fuzzy but comprehensible nexus of separated maps, only partly covering the terrain, but containing enough information for us to travel safely and arrive comfortably, at least most of the time.

The development from ordered certainty leaves us with a world uncertain and semantically ill defined, but in many ways it seems to describe the totality of human experience in an efficient way. Many of us regard its structure as logically unsatisfactory, epistemologically hazy, and ontologically open. What there exists cannot be inferred from the signs on the map. Ontology is not determined by even the most successful epistemology. This leaves the scene open to various models of so-called "reality". Nothing excludes even the most esoteric creeds and religions. What we know does not support all we think we know.

The situation is far from satisfactory, but it may, in the end, be the best world image available to our limited human intellect. If that is so, we have to be grateful for what we get. Chasing rainbows has never uncovered the treasures. But the display of colors is magnificent.

REFERENCES

[1] T. S. Eliot, *The Waste Land* (1921).
[2] R. Carnap, *Der logische Aufbau der Welt* (1928).
[3] J. R. Searle, *The Rediscovery of the Mind* (MIT Press, 1992).
[4] Bas C. van Fraassen, *The Scientific Image* (Clarendon Press, 1980).
[5] D. Bohm, *Wholeness and the Implicate Order* (Routledge, 1980).
[6] Niels Bohr, *Physical Review* **48** (1935), 696.
[7] E. Mach, *Space and Geometry* (Open Court, 1906).
[8] R. Penrose, *The Emperor's New Mind* (Oxford University Press, 1989).
[9] C. G. Jung, *Archetypes and the Collective Unconscious, Collected Works,* vol. ix/1, (Routledge & Kegan Paul, 1959).
[10] L. Wittgenstein, *Tractatus Logico-Philosophicus* (original German edn., 1921).
[11] S. Weinberg, *Dream of a Final Theory* (Vintage, 1993).
[12] P. A. Schilpp and L. E. Hahn, *The Philosophy of Georg Henrik von Wright,* The Library of Living Philosophers vol. xix, (Open Court, 1989).
[13] R. P. Feynman, *Theory of Fundamental Processes* (1961).
[14] L. Boltzmann, *Theoretical Physics and Philosophical Problems* (Reidel, 1974).
[15] P. Feyerabend, *Against Method* (Verso, 1978).
[16] C. G. Jung, *The Archtypes and the Collective Unconscious, Collected Works,* vol. ix/1, p. 32 (Routledge & Kegan Paul, 1959).
[17] C. G. Jung, *Aion, Collected Works,* vol. ix/2, p. 13 (Routledge & Kegan Paul, 1959).
[18] Andrew Wheatcroft, *Habsburg: Embodying Empire* (Penguin Books, 1996).
[19] R. Musil, *The Man Without Qualities* (Vintage, 1996) (original German edn., 1952 and later).
[20] Stefan Zweig, *The World of Yesterday* (University of Nebraska Press, 1964).
[21] A. Janik and S. E. Toulmin, *Wittgenstein's Vienna* (Simon and Schuster, 1973).
[22] Richard von Kraft-Ebing, *Psychopathia Sexualis* (Arcade Publishing, 1965).
[23] Harald Salfellne, *Franz Kafka and Prague* (Vitalis, 2003).
[24] Poul M. Møller, *En dansk students eventyr* (Det Schubotheske Forlag, 1896).
[25] Johannes Eduard Hohlenberg, *Søren Kierkegaard* (Octagon books, 1978).
[26] Alistair Hannay, *Kierkegaard: A Biography*(Cambridge University Press, 2001).
[27] F. Copleston, *Contemporary Philosophy: Studies of Logical Positivism and Existentialsim* (Continuum, 1956).

[28] S. Kierkegaard, *Concluding Unscientific Postscript* (1846).
[29] J. Honner, *The Description of Nature: Niels Bohr and the Philosophy of Quantum Physics* (Clarendon Press, 1987).
[30] D. Murdoch, *Niels Bohr's Philosophy of Physics* (Cambridge University Press, 1987).
[31] H. Folse, *The Philosophy of Niels Bohr* (North-Holland, 1985).
[32] A. Pais, *Niels Bohr's Times* (Oxford University Press, 1991).
[33] J. Faye, *Niels Bohr: His Heritage and Legacy*(Kluwer, 1991).
[34] A. Plotnitsky, *Reading Bohr: Physics and Philosophy* (Springer, 2006).
[35] *The Philosophical Writings of Niels Bohr*, vol. i. *Atomic Theory and the Description of Nature.*
[36] *The Philosophical Writings of Niels Bohr*, vol. ii. *Essays 1933–1957 on Atomic Physics and Human Knowledge.*
[37] *The Philosophical Writings of Niels Bohr*, vol. iii. *Essays 1958–1962 on Atomic Physics and Human Knowledge.*
[38] *The Philosophical Writings of Niels Bohr*, vol. iv. *Causality and Complementarity.*
[39] Niels Bohr, *Physics and the World*, ed. H. Feshbach, T. Matsui, and A. Oleson Harwood (1988).
[40] B. Russell, *An Inquiry into Meaning and Truth* (Allen and Unwin, 1978).
[41] L. Wittgenstein, *Tractatus Logico-Philosophicus* (originally published 1921).
[42] A. Petersen, *Quantum Physics and the Philosophical Tradition* (MIT Press, 1968).
[43] L. Rosenfeld, *Complementarity and Statistics*, i (Kongl. Norske Videnskaps Delskabs Förhandlinger, Bind **31**, 1958 Nr 9).
[44] L. Rosenfeld, *Complementarity and statistics*, ii (Kongl. Norske Videnskaps Delskabs Förhandlinger, Bind **31**, 1958 Nr 10).
[45] L. Rosenfeld, *Niels Bohr: An Essay Dedicated to Him on the Occation of his Sixtieth Birthday* (2nd Edn., North-Holland, 1961 and Nordita Publications, No. 57).
[46] A.I. Baz', Ya. B. Zeldovich, and A. M. Perelomov, *Scattering, Reactions and Decay in Nonrelativistic Quantum Mechanics* (Israel Program for Scientific Translations, 1969), p. 193.
[47] L. Rosenfeld, *Foundations of Quantum Theory and Complementarity*, *Nature*, **190** (1961), 384.
[48] L. Rosenfeld, *Complementarity*, *ICSU Review* **4** (1962), 42 and Nordita Publications, no. 66.
[49] L. Rosenfeld, *Niels Bohr's contribution to Epistemology*, *Physics Today*, **16**/10 (1963), 47.
[50] L. Rosenfeld,*The Epistemological Conflict between Einstein and Bohr*, *Z. Physik*, **171** (1963), 242 and Nordita Publications, no. 91.
[51] L. Rosenfeld, *The Measuring Process in Quantum Mechanics*, suppl. of *Progress in Theor. Phys*, (1965), p. 222 and Nordita Publications, no. 186.

[52] L. Rosenfeld, *The Structure of Quantum Theory, Revue de synthèse*, **3/49–52** (1968), 257 and Nordita Publications, No. 307.
[53] L. Rosenfeld, *Review of W. Heisenberg, Natural Law and the Structure of Matter, Nature*, **228** (1970), 479.
[54] L. Rosenfeld, *Unphilosophical Considerations on Causality in Physics, Perspectives in Quantum Theory, Essays in Honor of Alfred Landé* (MIT Press, 1971) and Nordita Publications, no. 434.
[55] W. James, *The Principles of Psychology* (1890).
[56] W. James, *The Will to Believe and Other Essays in Popular Philosophy* (1897).
[57] W. James, *The Varieties of Religious Experience: A Study in Human Nature* (1902).
[58] W. James, *Pragmatism: A New Name for Some Old Ways of Thinking* (1907).
[59] W. James, *A Pluralistic Universe* (1909).
[60] W. James, *The Meaning of Truth: A Sequel to "Pragmatism"* (1909).
[61] W. James, *Some Problems of Philosophy: A Beguinning to an Introduction to Philosophy* (1911).
[62] W. James, *Essays in Radical Empirism* (1912).
[63] W. A. Fock, *Kritik der Anschauungen Bohrs über die Quantenmechanik,* Sowjetwissenschaft: Naturwissenschaftlische Abtailung, **5** (1952), 123
[64] F. Engels, *Dialectics of Nature.*
[65] V. Lenin, *Materialism and Empirio-criticism.*
[66] L. R. Graham in *Niels Bohr, Physics and the World,* ed. H. Feshbach, T. Matsui and A. Oleson (Harwood Academic Publishers, 1985), p. 305.
[67] M. E. Omelyanovsky, *Dialectics in Modern Physics* (Progress Publishers, 1970).
[68] H. Weyl, *Mind and Nature: Selected Writings on Philosophy, Mathematics, and Physics* (Princeton University Press, 2009).
[69] A. N. Whitehead, *Concept of Nature* (Cambridge University Press, 1971).
[70] H. Poincaré, *Science and Hypothesis* (The Modern Library, 2001) (originally published 1905).
[71] H. Weyl, *Philosophy of Mathematics and Natural Science* (Princeton University Press, 2009) (original German published 1927).
[72] H. Weyl, *The Continuum* (Dover, 1994).
[73] Papers in P. Mancosu, *From Brouwer to Hilbert* (Oxford University Press, 1998).
[74] P. W. Bridgeman,*The Logic of Modern Physics* (MacMillan Company, 1927).
[75] A. I. Akhieser and R. V. Polovin, *Soviet Physics Uspekhi*, **15** (1973), 500.
[76] R. Monk, *Wittgenstein: The Duty of Genius* (Vintage, 1990).
[77] L. Wittgenstein, *Philosophical Investigations* (Blackwell, 1968).
[78] F. Waismann, *L. Wittgenstein and the Vienna Circle* (Blackwell, 1979).
[79] L. Wittgenstein, *Philosophical Grammar* (Blackwell, 1974).

[80] L. Wittgenstein, *Philosophical Occasions 1912–1951*(Hackett, 1993).
[81] L. Wittgenstein, *Philosophical Remarks* (Blackwell, 1975).
[82] L. Wittgenstein, *Remarks on Logical Form* ([80], p. 29).
[83] L. *Wittgenstein's Lectures 1930–1932* (Chicago, 1980).
[84] L. *Wittgenstein's Lectures 1932–1935* (Chicago, 1979).
[85] *Moore's L. Wittgenstein Lectures 1930–1933* ([80], p. 46).
[86] L. Wittgenstein, *Yellow Book* ([84], p. 43).
[87] L. Wittgenstein, *Foundations of Mathematics 1939* (Chicago, 1976).
[88] L. Wittgenstein, *The Blue and Brown Books* (Harper, 1958).
[89] L. Wittgenstein,*Philosophy for Mathematics* ([84], p. 205).
[90] L. Wittgenstein, *Zettel* (University of California Press, 1970).
[91] L. Wittgenstein, *Culture and Value* (Blackwell, 1980).
[92] L. Wittgenstein, *Remarks on Colour* (Blackwell, 1991).
[93] E. Stenius, *Wittgenstein's Tractatus* (Thoemmes, 1996).
[94] H.O. Mounce, *Wittgenstein's Tractatus, an Introduction* (Blackwell, 1981).
[95] Plato, *Dialogues.*
[96] G. E. Moore, *A Defence of Common Sense* (Contemporary British Philosophy, 1925).
[97] G. E. Moore, *Proof of an External World* (Proceedings of the British Academy, 1930).
[98] L. Wittgenstein, *On Certainty* (Blackwell, 1969).
[99] G. H. von Wright, *Wittgenstein* (Blackwell, 1982).
[100] M. Marion, *Wittgenstein, Finitism, and the Foundations of Mathematics* (Clarendon Press, 1998).
[101] H. Putnam, *Pragmatism* (Blackwell, 1995).
[102] W. van Orman Quine,*From a Logical Point of View* (Harper, 1953).
[103] E. Husserl, *The Crisis of European Sciences* (Northwestern, 1970).
[104] L. Wittgenstein, *Notebooks 1914–1916* (Blackwell, 1979).

INDEX

Anscombe Elizabeth, 131
assertibility, 76, 77

behaviorism, 20
binary description, 186
  yes-no question, 110

Carnap, 154
Church-Turing thesis, 169
common language, 2
Conferences, 64
conservation laws, 60
Copenhagen interpretation, 22, 48
  positivism
    pragmatism, 77
correspondence, 61
correspondence rules, *see* projection
counterfactual, 64, 65, 194

dialectic, 38, 40, 50
  Hegel, 39
  method, 189
    materialism, 18
    Socrates, 38
dialectical materialism, 83
doubt, 2, 160, 161, 163

eidic science
  epoché, 200
Einstein-Podolsky-Rosen, *see* Solvay
EKLIPTIKA, 55
epistemology, 4, 6, 12, 50, 69
excluded middle, *see* intuitionism
existentialism, 5, 23, 30, 36
  European, 30
extensive abstraction, 113

forms of life, *see* life world
forms of perception, 58, 59
Freud, 26

God, 6, 33, 151, 208
  existence, 178
  metaphysical attributes
    moral attributes, 6
  underdetermined, 211
Goethe, 33, 34, 40, 172

Heisenberg, 77
hypothesis, 57, *see* proposition

indeterminity, 67, *see* uncertainty
instrumentalism, 196
intuitionism, 166

Jew, 134, 138, 154
Jewish, 130
Jung, 20

knowledge, 7, 50, 87, 188
  antiquity, 32
  belief, 54
  empirical, 115
  foundations, 204
  Jews, 25
  Marx, 68
Kraus, 29

language, 21, 62, 82,
    173, 192
  approach to reality, 23
  common, 2
language game, 2
leap of faith, 54
Lenin, 8
life world, 201, 202
logical multiplicity, 174
logical space, 150
  quantum system, 180

Mach, 132
map
  intention of, 176
  of reality, 150
  projection, 173
Marxist, 68
measurement
  quantum, 51, 58
  unlearning, 63
measurements
  Bohr measurements, 63
  Bohr's description, 63
  complementary, 65
  quantum, 51
metaphysics, 12

model
  physics
    philosophy, 170
Musil
  "Kakania", 25

Nachlass, 146, 161, 171, 180
nonlocality
  quantum
    probability, 64
NORDITA, 69

ontological, 57
ontology, 100

Penrose, 9
phenomenon, 49, 51, 164
  closed, 63
  life, 71
  natural, 15
  physical, 2
  quantum, 62
positivism
  Vienna Circle, 51
possible world, 87, 92, 115
pragmatism, 4, 5, 52, 77
  American, 76
  philosophy of science, 214
  tenets of, 197
preparation, 118
private language, 159
probability
  confirmed
    verified, 190
  hypothesis, 190
  objective, 188
projection, 173
provability, 104

quantum
  knowledge, 15
quantum system
  preparation
    entanglement, 118
  primary properties
    secondary properties, 63

reality, 2, 60, 80, 83, 84, 94, 99, 100, 189, 206, 215
  correspondence, 56
  empirical
    binary description, 110
    theoretical, 12
  exist, 8
  external, 9
  features of, 59
  internal, 21
  linguistic, 11
  logical construct, 108
  maps of, 169
  mirror of, 127, 199
  of atoms, 60
  Platonic, 101
  quantum description, 116
  ultimate, 6
  world of facts, 187
Rees Rush, 131

scienticism, 202
set, 1, 102
  theory
    finite, 102
Solvay, 49
  Einstein-Podolsky-Rosen, 65
states of affairs, 65, 149
subconscious, 20

Tarski, 13
truth table, 153
Turing, 8

uncertainty, 60, *see* indeterminity, 81, 213

Vienna
  fin-de-siècle
    cafe culture, 4
Vienna Circle, 154
von Wright, 131

warranted assertibility, 76
world, 87, 197
  absurd, 15
  construction, 18
  empirical, 2
  everyday, 68
  image, 215
  logically complete, 165
  material, 96
  of facts, 187
  possible, 87
  real, 8
  well-ordered, 12
  whole, 115

Zweig, 26